Holt Algebra 1

Chapter 6 Resource Book

HOLT, RINEHART AND WINSTON

A Harcourt Education Company

Orlando • Austin • New York • San Diego • London

Contents

Holt Algebra 1

Date ___________________

Dear Family,

In Chapter 6, your child will study systems of linear equations and inequalities.

A **system of linear equations** is a set of two or more linear equations containing two or more variables. A **solution to the system** is a point that makes all of the equations true.

Graphically, a solution to a system of two linear equations in two variables is the point where the two lines intersect.

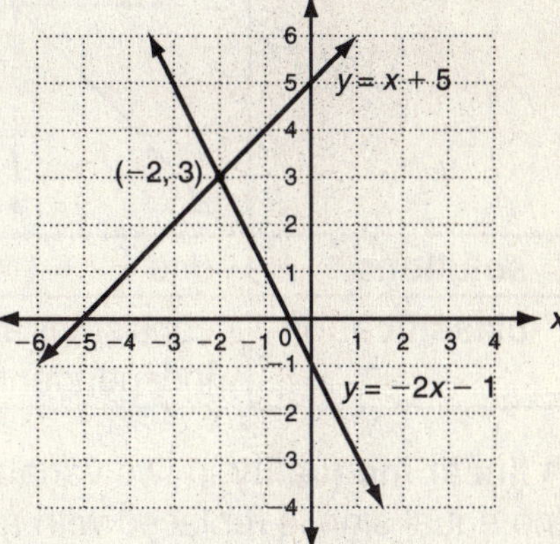

linear system:	**solution to the system:**
$\begin{cases} y = -2x - 1 \\ y = x + 5 \end{cases}$	$(-2, 3)$

check:

$y = -2x - 1$	
3	$-2(-2) - 1$
3	$3\checkmark$

$y = x + 5$	
3	$-2 + 5$
3	$3\checkmark$

One algebraic method of solving a system is called **substitution**.

system: $\begin{cases} 2x + y = -4 \\ x + y = -7 \end{cases}$

$2x + y = -4$
$y = -2x - 4$

Step 1: Solve one equation for one variable (if necessary).

$x + y = -7$

Step 2: <u>Substitute</u> the expression into the other equation.

$x + (-2x - 4) = -7$
$-x - 4 = -7$
$x = 3$

Step 3: Solve for the value of the first variable.

$x + y = -7$
$3 + y = -7$
$y = -10$

Step 4: Substitute the value into one of the original equations and solve for the value of the other variable.

$(3, -10)$

Step 5: Write the values as an ordered pair.

Another algebraic method is called **elimination**.

$2x + y = -4$

Step 1: Write the system so that like terms are aligned.

$-(x + y = -7)$
$\overline{1x + 0 = 3}$
$x = 3$

Step 2: Add or subtract multiples of the equations in order to <u>eliminate</u> one of the variables...

... and solve for the value of the remaining variable.

$3 + y = -7$
$y = -10$

Step 3: Substitute and solve for the other variable.

$(3, -10)$

Step 4: Write the values as an ordered pair.

Holt Algebra 1

Not all linear systems have exactly one solution. If the lines coincide, then there are infinitely many solutions. If the lines are parallel and never intersect, then there is no solution. Based on their solutions, systems are classified as **consistent** (at least one), **inconsistent** (none), **independent** (exactly one), or **dependent** (infinitely many).

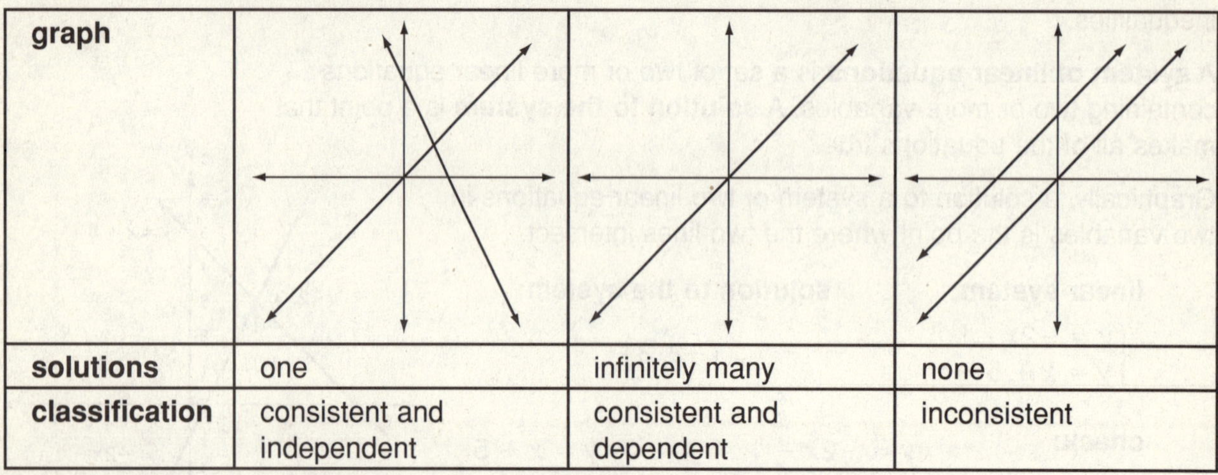

graph			
solutions	one	infinitely many	none
classification	consistent and independent	consistent and dependent	inconsistent

A **linear inequality** in two variables is similar to a linear equation, but the equal sign is replaced with an inequality ($>$, $<$, $\geq$, or $\leq$). Similarly, a **solution of a linear inequality** is any ordered pair that makes the inequality true.

Linear inequalities are graphed first with a dashed or solid boundary line. Then you shade above or below the line to show all of the points that satisfy the inequality. The graph of a linear inequality is called a **half-plane**.

Graphing a Linear Inequality Solved for y (slope-intercept form)		
Boundary	$\leq$ or $\leq$ (equal to)	Solid
	$>$ or $<$ (not equal to)	Dashed
Shading	$>$ or $\geq$ (greater than)	Above
	$<$ or $\geq$ (less than)	Below

linear inequality: $y < x + 1$

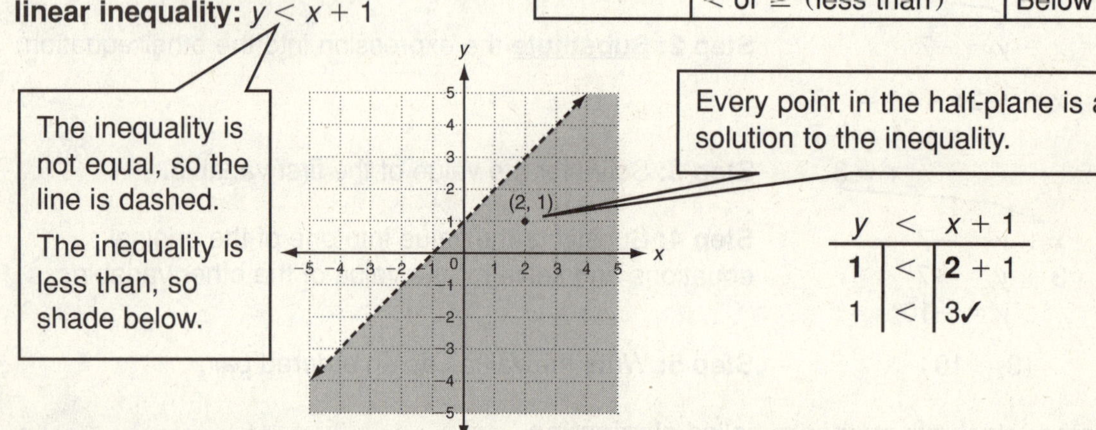

A **system of linear inequalities** is a set of two or more linear inequalities containing two or more variables. The **solution of a system of linear inequalities** consists of all the ordered pairs that satisfy all of the linear inequalities. Graphically, the solution occurs where the half-planes of the linear inequalities intersect, or the overlapping shaded regions.

For additional resources, visit go.hrw.com and enter the keyword MA7 Parent.

Holt Algebra 1

LESSON 6-1

Practice A
Solving Systems by Graphing

Complete the steps to determine whether the ordered pair is a solution of the given system. Circle ✓ or ✗ for each equation. Then, write *is* or *is not* to complete the sentence.

1. $(2, 4)$; $\begin{cases} x - y = -2 \\ 2x + y = 6 \end{cases}$

2. $(1, -2)$; $\begin{cases} 2x + y = 0 \\ x + 4y = -7 \end{cases}$

$x - y = -2$	
$(2) - (4)$	-2
_____	-2

✓ or ✗

$2x + y = 6$	
$2(2) + (\_\_\_)$	6
$(\_\_\_) + (\_\_\_)$	6
$(\_\_\_)$	6

✓ or ✗

$2x + y = 0$	
$2(1) + (-2)$	0
$(\_\_\_) + (-2)$	0
$(\_\_\_)$	0

✓ or ✗

$x + 4y = -7$	
$(1) + 4(\_\_\_)$	-7
$(\_\_\_) + (\_\_\_)$	-7
$(\_\_\_)$	-7

✓ or ✗

$(2, 4)$ _________ a solution of the system.

$(1, -2)$ _________ a solution of the system.

Solve each system by graphing. One of the lines has been graphed for you.

3. $\begin{cases} y = 3x - 5 \\ y = x - 3 \end{cases}$ Solution: _____________

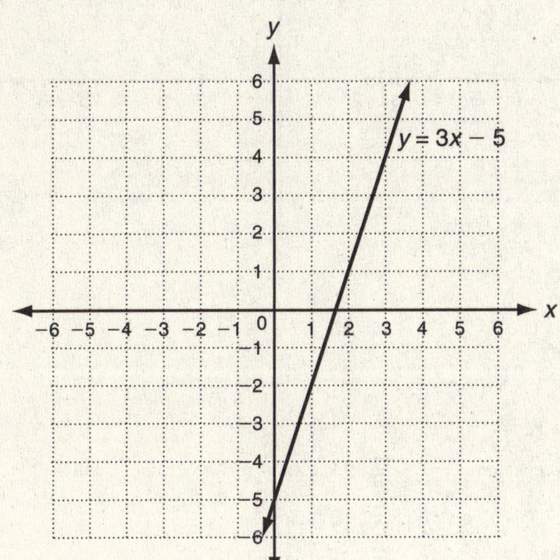

4. $\begin{cases} y = x + 7 \\ y = -2x - 2 \end{cases}$ Solution: _____________

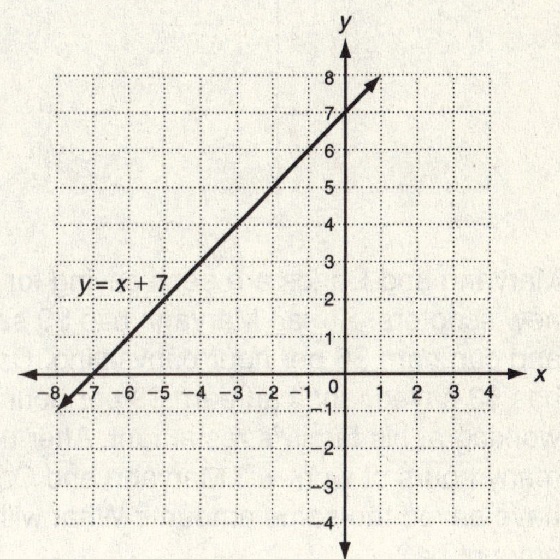

5. The Science Club needs to rent a bus for a field trip. Main Street Buses charges a $40 rental fee, plus $2 per mile. County Bus Line charges a $20 rental fee, plus $3 per mile. For what number of miles will the total charge be the same? What will that charge be?

____

3

Practice B
Solving Systems by Graphing

Tell whether the ordered pair is a solution of the given system.

1. $(3, 1)$; $\begin{cases} x + 3y = 6 \\ 4x - 5y = 7 \end{cases}$ __________

2. $(6, -2)$; $\begin{cases} 3x - 2y = 14 \\ 5x - y = 32 \end{cases}$ __________

$x + 3y = 6$	$4x - 5y = 7$		$3x - 2y = 14$	$5x - y = 32$

Solve each system by graphing. Check your answer.

3. $\begin{cases} y = x + 4 \\ y = -2x + 1 \end{cases}$ Solution: __________

4. $\begin{cases} y = x + 6 \\ y = -3x + 6 \end{cases}$ Solution: __________

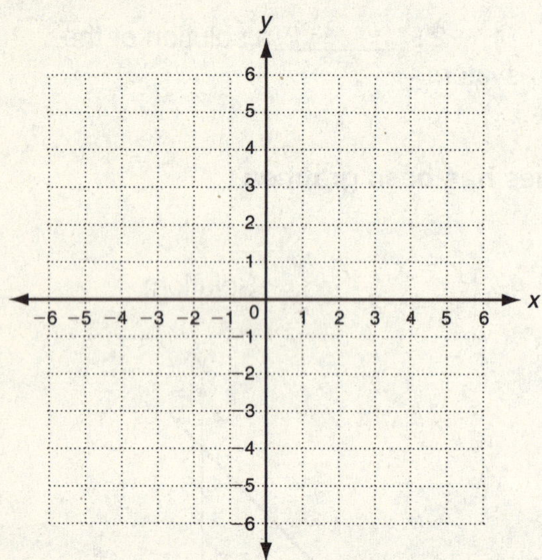

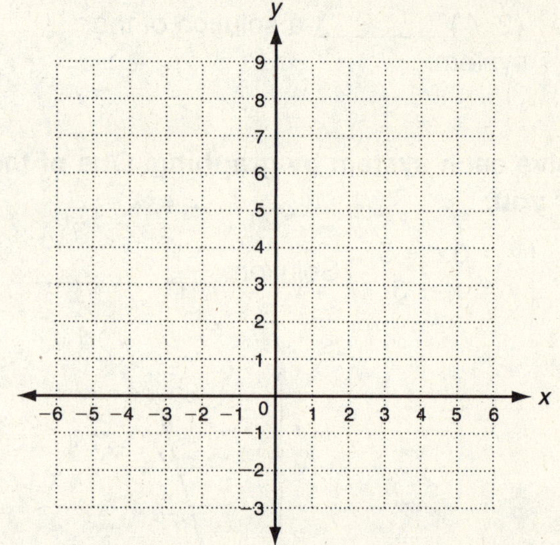

5. Maryann and Carlos are each saving for new scooters. So far, Maryann has $9 saved, and can earn $6 per hour babysitting. Carlos has $3 saved, and can earn $9 per hour working at his family's restaurant. After how many hours of work will Maryann and Carlos have saved the same amount? What will that amount be?

____

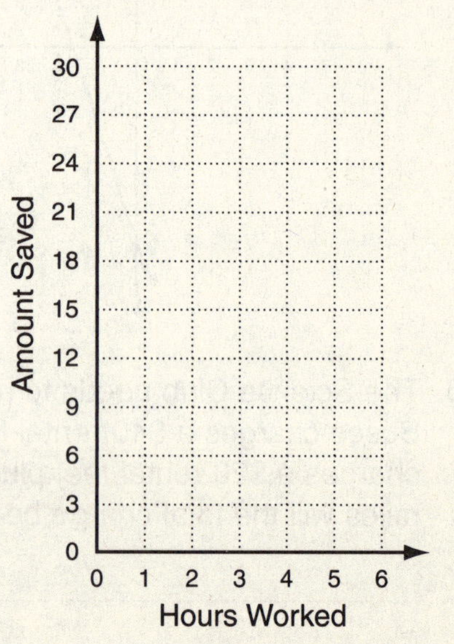

Holt Algebra 1

<table><tr><td>**LESSON**
6-1</td><td>## Practice C
Solving Systems by Graphing</td></tr></table>

Tell whether the ordered pair is a solution of the given equation.

1. $(6, -2)$; $\begin{cases} 2x - y = 14 \\ x + 4y = -2 \end{cases}$

2. $(4, 0)$; $\begin{cases} x - 2y = 4 \\ -x + y = -8 \end{cases}$

3. $(-6, -2)$; $\begin{cases} 2x - y = -10 \\ -x + y = 4 \end{cases}$

______________   ______________   ______________

Solve each system by graphing.

4. $\begin{cases} y = 2x + 4 \\ y = -x + 7 \end{cases}$

5. $\begin{cases} y = 2x - 6 \\ y = 3x - 8 \end{cases}$

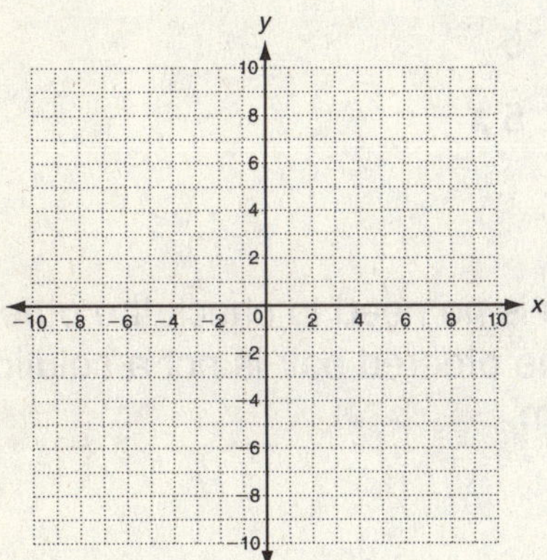

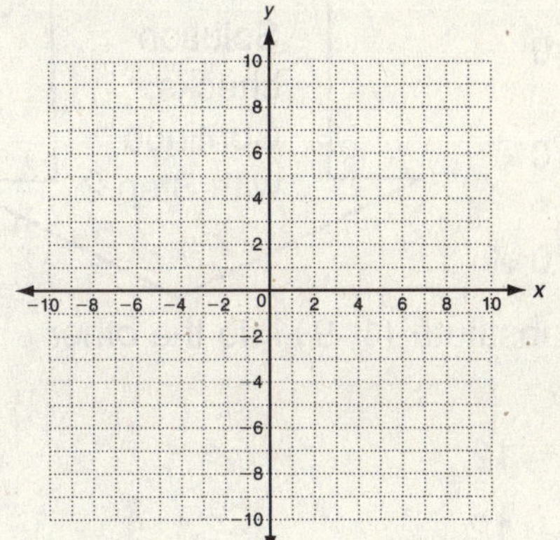

Solution: ________________

Solution: ________________

6. $\begin{cases} x + y = -2 \\ y = 4x - 7 \end{cases}$

7. $\begin{cases} x = y + 2 \\ 2x = y \end{cases}$

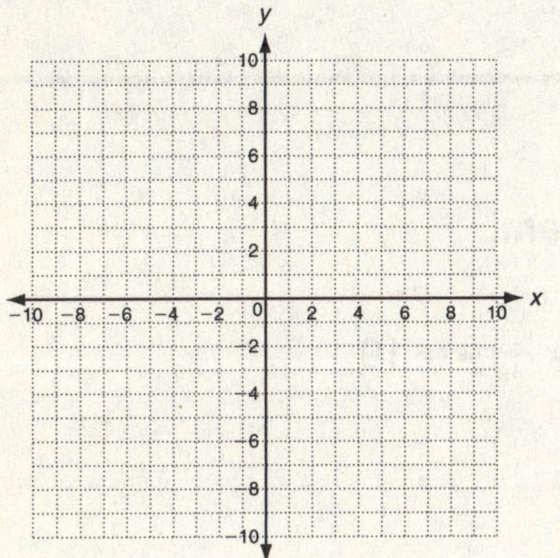

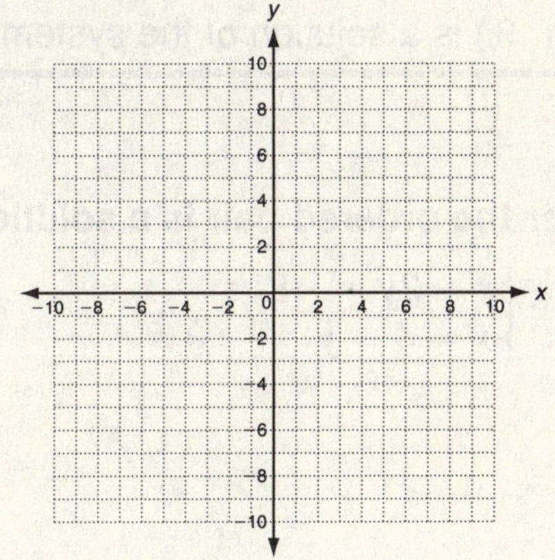

Solution: ________________

Solution: ________________

Use a graphing calculator to solve.

8. To sell an item in an online auction, WebAuctions charges a $5 listing fee plus 10% of the final selling price. AuctionsOnline charges a $3 listing fee plus 15% of the final selling price. For what final selling price do both companies charge the same amount? What will that amount be? ________________________

Holt Algebra 1

Reteach

LESSON 6-1

Solving Systems by Graphing

You have checked to see if an ordered pair was a solution of an equation. Now you will check to see if an ordered pair is a solution of a system of equations.

Tell whether $(1, 9)$ is a solution of
$$\begin{cases} x + y = 10 \\ 3x + y = 12 \end{cases}$$

Step 1: Substitute $(1, 9)$ into one of the equations.

$(1, 9)$ means that $x = 1$ and $y = 9$.

$x + y \overset{?}{=} 10$

$1 + 9 \overset{?}{=} 10$

$10 \overset{?}{=} 10$ ✓

Solution checks. Continue with Step 2

Step 2: Substitute $(1, 9)$ into the other equation.

$3x + y = 12$

$3(1) + 9 \overset{?}{=} 12$

$3 + 9 \overset{?}{=} 12$

$12 \overset{?}{=} 12$ ✓

The ordered pair makes both equations true. So $(1, 9)$ is a solution of the system.

Tell whether $(2, -3)$ is a solution of
$$\begin{cases} x + y = 5 \\ 2x + 5y = -11 \end{cases}$$

Step 1: Substitute $(2, -3)$ into one of the equations.

$x + y = 5$

$2 + -3 \overset{?}{=} 5$

$-1 \overset{?}{=} 5$ ✗

Stop! There is no need to check the other equation. The ordered pair is not a solution of the system.

Tell whether the ordered pair is a solution of the given system.

1. $(0, -4)$; $\begin{cases} x + 2y = -8 \\ x = 4 + y \end{cases}$

2. $(2, 5)$; $\begin{cases} x + y = 7 \\ 3x + y = 10 \end{cases}$

____

____

3. $(-3, 1)$; $\begin{cases} 2x + y = 5 \\ x + 3y = -6 \end{cases}$

4. $(-3, 9)$; $\begin{cases} y = x + 12 \\ y = -3x \end{cases}$

____

____

Holt Algebra 1

Reteach
Solving Systems by Graphing (continued)

Graph to check if $(5, 7)$ is a solution of $\begin{cases} y = x + 2 \\ y = 2x + 3 \end{cases}$.

If $(5, 7)$ is not the solution, find the solution from the graph.

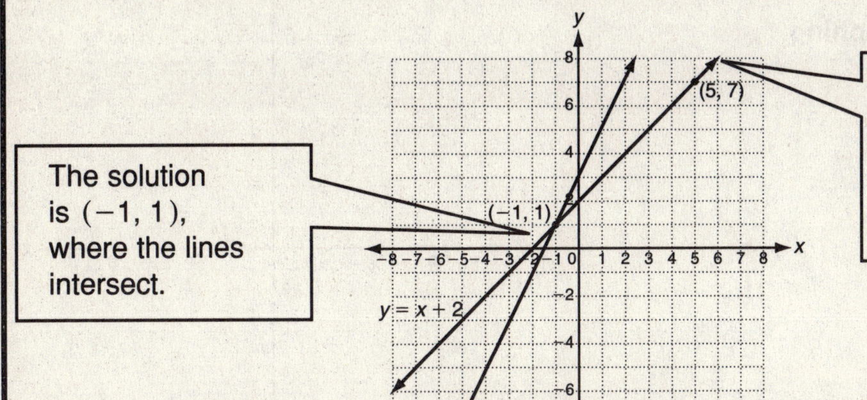

Find the solution of each system of equations graphed below.

5.

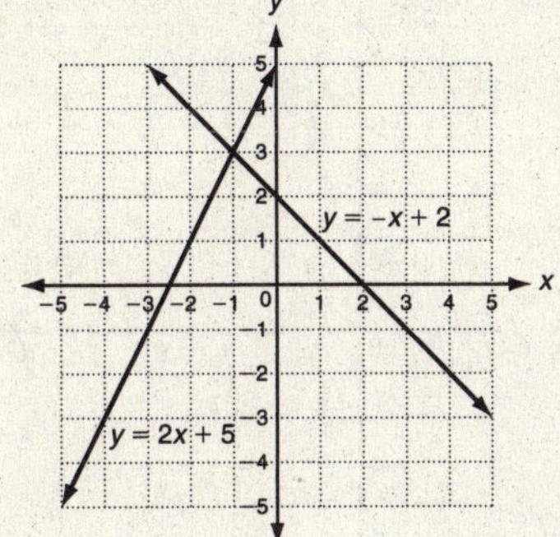

____

6.

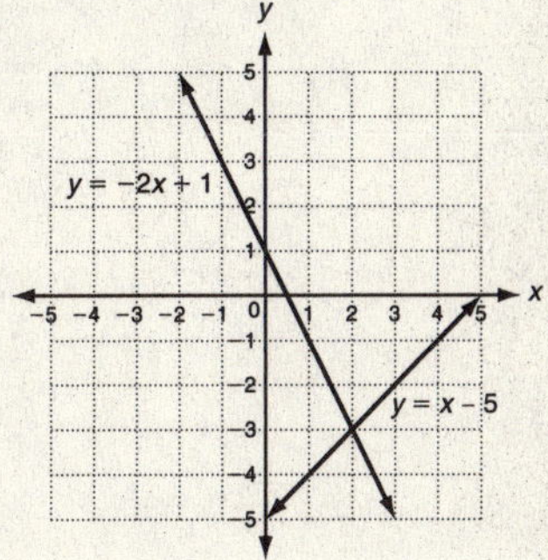

____

Solve each system by graphing.

7. $\begin{cases} y = -3 \\ y = x + 2 \end{cases}$

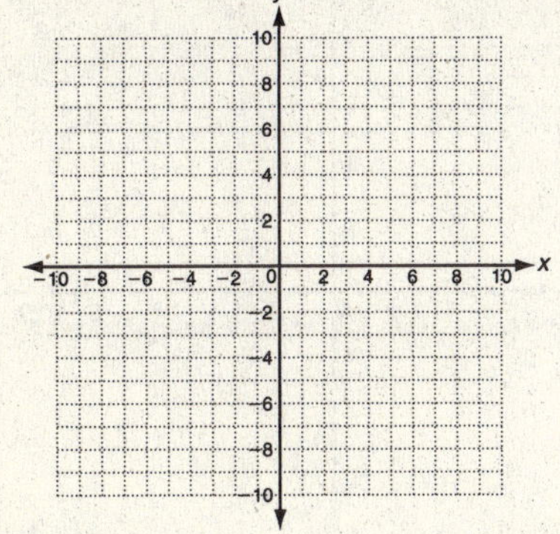

8. $\begin{cases} y = x - 6 \\ y = -x \end{cases}$

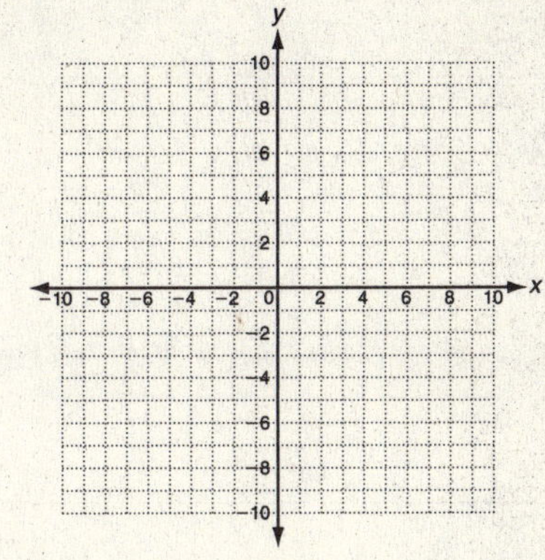

_______________________________   _______________________________

Holt Algebra 1

Challenge
Solving Systems by Graphing

Sometimes one or both equations in a system will be non-linear. The solutions to these systems will still be where the graphs intersect. This can happen in more than one place. Recall that you can graph any equation by generating and plotting ordered pairs.

Solve each system by graphing.

1. $\begin{cases} y = x^2 - 4 \\ y = x + 2 \end{cases}$

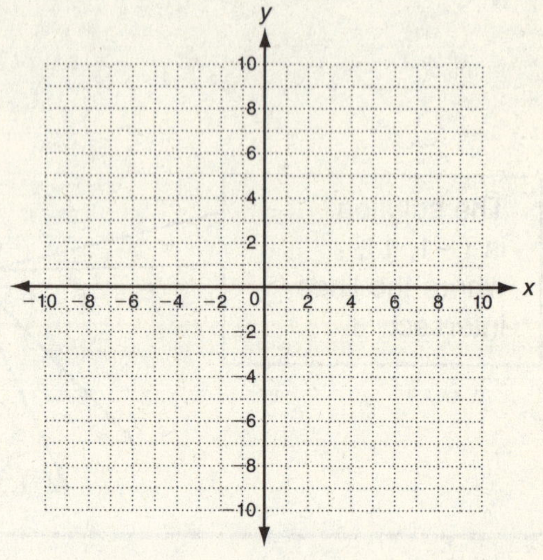

2. $\begin{cases} y = -x^2 + 5 \\ y = x^2 - 5 \end{cases}$

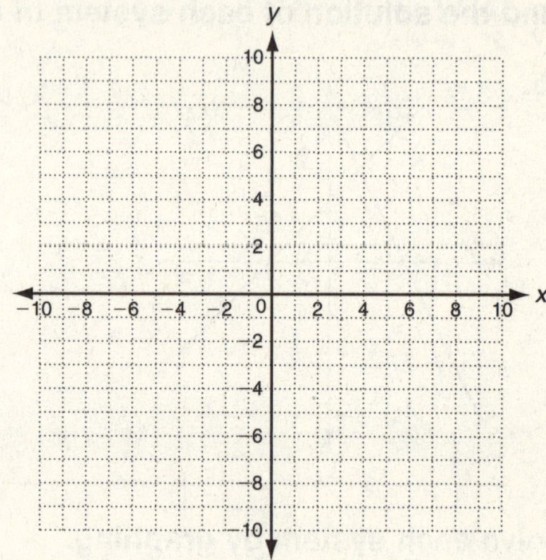

3. $\begin{cases} y = |x + 1| \\ y = \frac{1}{2}x + 3 \end{cases}$

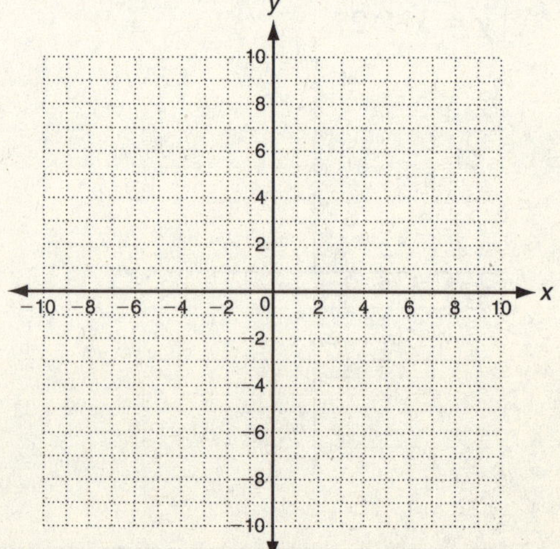

Holt Algebra 1

LESSON 6-1 Problem Solving
Solving Systems by Graphing

Write the correct answer.

1. Mr. Malone is putting money in two savings accounts. Account A started with $200 and Account B started with $300. Mr. Malone deposits $15 in Account A and $10 in Account B each month. In how many months will the accounts have the same balance? What will that balance be?

2. Tom currently has 5 comic books in his collection and has subscribed to receive 5 new comic books each month. His uncle has 145 comic books, but sends 5 to each of his 3 nieces each month. In how many months will they have the same number of comic books? How many books will that be?

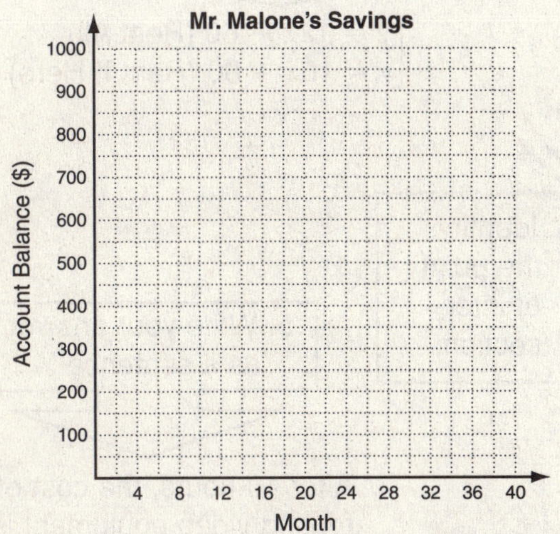

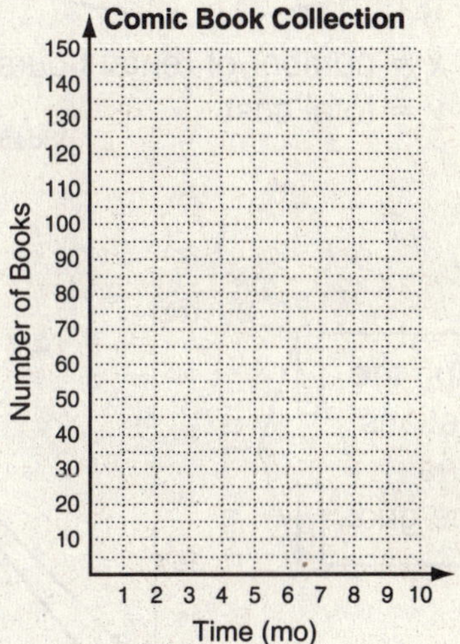

The graph below compares the heights of two trees. Use the graph to answer questions 3–5. Select the best answer.

3. How many years after planting will the trees be the same height?

   A 1 years   C 4 years

   B 2 years   D 6 years

4. Which system of equations is represented by the graph?

   F $\begin{cases} y = x + 2 \\ y = 0.5x + 2 \end{cases}$ H $\begin{cases} y = 2x + 4 \\ y = x + 4 \end{cases}$

   G $\begin{cases} y = x + 2 \\ y = 0.5x + 4 \end{cases}$ J $\begin{cases} y = 4x - 2 \\ y = 2x + 2 \end{cases}$

5. How fast does the tree that started at 2 feet tall grow?

   A 0.5 ft/yr   C 1.5 ft/yr

   B 1 ft/yr   D 2 ft/yr

6. How fast does the tree that started at 4 feet tall grow?

   F 0.5 ft/yr   H 1.5 ft/yr

   G 1 ft/yr   J 2 ft/yr

Holt Algebra 1

<table>
<tr><td>LESSON
6-1</td><td>

Reading Strategies
Follow a Procedure
</td></tr>
</table>

In the example below, the steps show how to set up and solve a system of linear equations.

A teacher needs to rent video equipment. Rent All charges $60 plus $12 for each hour. Rent It Here charges $80 plus $10 per hour. After how many hours is the cost of renting video equipment the same from both companies?

1. Define variables for the unknowns.	2. Write a system of linear equations.

- Let x = number of rental hours
- Let y = total cost

$$\begin{cases} y = 12x + 60 \text{ (Rent All)} \\ y = 10x + 80 \text{ (Rent It Here)} \end{cases}$$

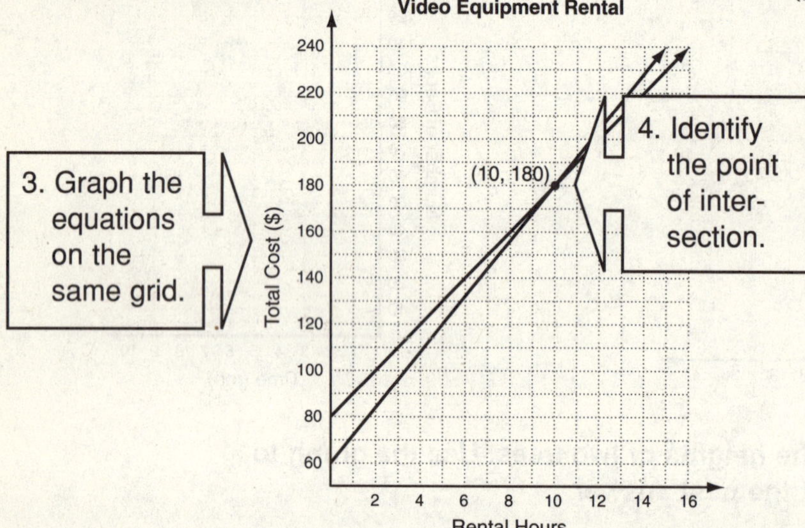

3. Graph the equations on the same grid.

4. Identify the point of intersection.

5. Write your answer as a sentence.

After 10 hours, the cost of renting video equipment is $180 from both companies.

Derek is going to have a party catered. Good Eats charges $120 plus $10 per person. Food Fare charges $150 plus $8 per person. Complete the following to find the number of people for which the total cost is the same for both catering companies.

1. Define the variables. Let x = _____________________ , Let y = _____________________

2. Write a system of linear equations. $\begin{cases} \underline{\hspace{3cm}} \\ \underline{\hspace{3cm}} \end{cases}$

3. Graph the equations.

4. What is the point of intersection? _____________________

5. Write your answer. _____________________

____

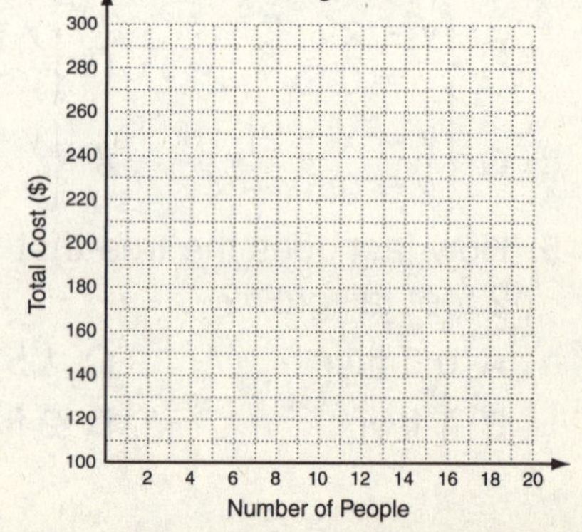

Holt Algebra 1

Practice A
Solving Systems by Substitution

Fill in the blanks to solve each system by substitution.

1. $\begin{cases} y = 3x \\ y = x + 4 \end{cases}$

Substitute ___________ for y in the second equation.

_______ $= x + 4$

$-$___ $-x$

_______ $= 4$

$\div$ ____ $\div$ ____

_______ $=$ _______

Since $x =$ ___, substitute ___ for x in one of the equations to find the value of y:

$y = 3x$

$y = 3\big($___$\big)$

$y =$ _______

Solution: $\big($_______ , _______$\big)$

2. $\begin{cases} 3x + y = 25 \\ y = x - 3 \end{cases}$

Substitute ___________ for y in the first equation.

$3x + \big($_______$\big) = 25$

_________ $- 3 = 25$

$\quad\quad +3 \quad +3$

_______ $= 28$

$\div$ ____ $\div$ ____

_______ $=$ _______

Since $x =$ ___, substitute ___ for x in one of the equations to find the value of y:

$y = x - 3$

$y =$ ___ $- 3$

$y =$ _______

Solution: $\big($_______ , _______$\big)$

Solve each system by substitution. Check your answer.

3. $\begin{cases} y = 4x \\ y = 2x + 6 \end{cases}$

____

4. $\begin{cases} y = x - 2 \\ 2x + y = 4 \end{cases}$

____

5. $\begin{cases} 2x + y = -1 \\ -x + y = -7 \end{cases}$

____

6. A professional organizer charges a $45 consultation fee, plus $20 per hour. Her competitor charges a $30 consultation fee, plus $26 per hour.

   a. Write a system of equations to represent the situation.

   ____

   b. Solve the system by substitution. For what number of hours will the total charge be the same? _____________________

   c. What will that charge be? _____________________________

Holt Algebra 1

Practice B
Solving Systems by Substitution

Solve each system by substitution. Check your answer.

1. $\begin{cases} y = x - 2 \\ y = 4x + 1 \end{cases}$

2. $\begin{cases} y = x - 4 \\ y = -x + 2 \end{cases}$

3. $\begin{cases} y = 3x + 1 \\ y = 5x - 3 \end{cases}$

____

____

____

4. $\begin{cases} 2x - y = 6 \\ x + y = -3 \end{cases}$

5. $\begin{cases} 2x + y = 8 \\ y = x - 7 \end{cases}$

6. $\begin{cases} 2x + 3y = 0 \\ x + 2y = -1 \end{cases}$

____

____

____

7. $\begin{cases} 3x - 2y = 7 \\ x + 3y = -5 \end{cases}$

8. $\begin{cases} -2x + y = 0 \\ 5x + 3y = -11 \end{cases}$

9. $\begin{cases} \frac{1}{2}x + \frac{1}{3}y = 5 \\ \frac{1}{4}x + y = 10 \end{cases}$

____

____

____

Write a system of equations to represent the situation. Then, solve the system by substitution.

10. The length of a rectangle is 3 more than its width. The perimeter of the rectangle is 58 cm. What are the rectangle's dimensions?

____

11. Carla and Benicio work in a men's clothing store. They earn commission from each suit and each pair of shoes they sell. For selling 3 suits and one pair of shoes, Carla has earned $47 in commission. For selling 7 suits and 2 pairs of shoes, Benicio has earned $107 in commission. How much do the salespeople earn for the sale of a suit? for the sale of a pair of shoes?

____

Holt Algebra 1

Practice C
Solving Systems by Substitution

Solve each system by substitution. Check your answer.

1. $\begin{cases} y = x - 6 \\ y = 2x - 8 \end{cases}$

2. $\begin{cases} y = x - 4 \\ y = -2x + 5 \end{cases}$

3. $\begin{cases} y = 2x + 8 \\ y = 8 \end{cases}$

_________________ _________________ _________________

4. $\begin{cases} x + y = -3 \\ x - y = -1 \end{cases}$

5. $\begin{cases} 4x + 2y = -2 \\ y = 6x - 5 \end{cases}$

6. $\begin{cases} \frac{1}{2}x = y - 1 \\ \frac{1}{3}x = y \end{cases}$

_________________ _________________ _________________

7. $\begin{cases} 4x + y = 3.4 \\ x = 4y \end{cases}$

8. $\begin{cases} x = -\frac{1}{4}y + 5 \\ 3x + 2y = 0 \end{cases}$

9. $\begin{cases} 3x = -y + 10 \\ 2x + 3y = -12 \end{cases}$

_________________ _________________ _________________

Solve.

10. The sum of two numbers is 39. The greater number is 5 less than 3 times the smaller number. What are the two numbers?

____

11. Adam has dimes and quarters in his pocket. There are 28 coins in all, and they are worth $4.60 altogether. How many quarters does Adam have?

____

12. At a pet store, Rhonda paid $11.50 for 3 dog bowls and 4 bones. Kelly paid $13 for 2 dog bowls and 8 bones. How much will Andrew pay for 4 dog bowls and 2 bones?

____

Holt Algebra 1

Reteach
Solving Systems by Substitution

You can use substitution to solve a system of equations if one of the equations is already solved for a variable.

Solve $\begin{cases} y = x + 2 \\ 3x + y = 10 \end{cases}$

Step 1: Choose the equation to use as the substitute.

> Use the first equation $y = x + 2$ because it is already solved for a variable.

Step 2: Solve by substitution.

$$\boxed{x + 2}$$

$$3x + y = 10$$

$3x + (x + 2) = 10$ *Substitute $x + 2$ for y.*

$\quad 4x + 2 = 10$ *Combine like terms.*

$$\underline{\quad -2 \quad -2}$$

$$4x = 8$$

$$\frac{4x}{4} = \frac{8}{4}$$

$$x = 2$$

Step 3: Now substitute $x = 2$ back into one of the original equations to find the value of y.

$$y = x + 2$$

$$y = 2 + 2$$

$$y = 4$$

The solution is $(2, 4)$.

Check:

Substitute $(2, 4)$ into both equations.

$y = x + 2$	$3x + y = 10$
$4 \overset{?}{=} 2 + 2$	$3(2) + 4 \overset{?}{=} 10$
$4 \overset{?}{=} 4 \checkmark$	$6 + 4 \overset{?}{=} 10$
	$10 \overset{?}{=} 10 \checkmark$

Solve each system by substitution. Check your answer.

1. $\begin{cases} x = y - 1 \\ x + 2y = 8 \end{cases}$

2. $\begin{cases} y = x + 2 \\ y = 2x - 5 \end{cases}$

3. $\begin{cases} y = x + 5 \\ 3x + y = -11 \end{cases}$

4. $\begin{cases} x = y + 10 \\ x = 2y + 3 \end{cases}$

Holt Algebra 1

LESSON 6-2

Reteach
Solving Systems by Substitution (continued)

You may need to solve one of the equations for a variable before solving with substitution.

Solve $\begin{cases} y - x = 4 \\ 2x + 3y = 27. \end{cases}$

Step 1: Solve the first equation for y.

$$y - x = 4$$

$$\underline{+x \quad +x}$$

$$y = x + 4$$

Step 2: Solve by substitution.

$$\boxed{x + 4}$$

$$2x + 3y = 27$$

$2x + 3(x + 4) = 27$ *Substitute x + 4 for y.*

$2x + 3x + 12 = 27$ *Distribute.*

$5x + 12 = 27$ *Combine like terms.*

$$\underline{-12 \quad -12}$$

$$5x = 15$$

$$\frac{5x}{5} = \frac{15}{5}$$

$$x = 3$$

Step 3: Now substitute $x = 3$ back into one of the original equations to find the value of y.

$$y - x = 4$$

$$y - 3 = 4$$

$$\underline{+3 \quad +3}$$

$$y = 7$$

The solution is $(3, 7)$.

Check:

Substitute $(3, 7)$ into both equations.

$y - x = 4$ $2x + 3y = 27$

$7 - 3 \overset{?}{=} 4$ $2(3) + 3(7) \overset{?}{=} 27$

$4 \overset{?}{=} 4 \checkmark$ $6 + 21 \overset{?}{=} 27$

                    $27 \overset{?}{=} 27 \checkmark$

Solve each system by substitution. Check your answer.

5. $\begin{cases} x - y = -3 \\ 2x + y = 12 \end{cases}$

6. $\begin{cases} y - x = 8 \\ 5x + 2y = 9 \end{cases}$

Holt Algebra 1

Challenge
Three Equations in Three Variables

To solve a system with two variables, you must have two equations. To solve a system in three variables, you must have three equations.

Use substitution to find the values of *x*, *y*, and *z*.

1. $\begin{cases} 2x + 4y + 3z = 14 \\ y + 3z = 11 \\ z = 4 \end{cases}$

2. $\begin{cases} 4x + y + 3z = -9 \\ 2y - 5z = 23 \\ z = -3 \end{cases}$

____

3. $\begin{cases} 2x + 3y + 2z = 18 \\ -4x + y - z = 5 \\ 2z = -10 \end{cases}$

4. $\begin{cases} x + y + z = 15 \\ 2x - y + 3z = -5 \\ -x - y + z = -7 \end{cases}$

[Hint: Solve for *z* and get two
equations to have *x* and *y* only.]

_______________________ _______________________

Holt Algebra 1

Problem Solving

LESSON 6-2

Solving Systems by Substitution

Write the correct answer.

1. Maribel has $1.25 in her pocket. The money is in quarters and dimes. There are a total of 8 coins. How many quarters and dimes does Maribel have in her pocket?

   ____

   ____

2. Fabulously Fit offers memberships for $35 per month plus a $50 enrollment fee. The Fitness Studio offers memberships for $40 per month plus a $35 enrollment fee. In how many months will the fitness clubs cost the same? What will the cost be?

   ____

3. Vong grilled 21 burgers at a block party. He grilled the same number of pounds of turkey burgers as hamburgers. Each turkey burger weighed $\frac{1}{4}$ pound and each hamburger weighed $\frac{1}{3}$ pound. How many of each did Vong grill?

   ____

   ____

4. Kate bought 3 used CDs and 1 used DVD at the bookstore. Her friend Joel bought 2 used CDs and 2 used DVDs at the same store. If Kate spent $20 and Joel spent $22, determine the cost of a used CD and a used DVD.

   ____

   ____

Use the chart below to answer questions 5–8. Select the best answer.
The chart compares the quotes that the Masons received from four different flooring contractors to tear out and replace a floor.

5. Which expression shows the total cost if the work is done by Dad's Floors?

   A $8 + 150x$ C $150(8x)$

   B $150 + 8x$ D $158x$

6. How many square feet would the Masons need to have installed to make the total cost of V.I.P. Inc. the same as the total cost of Floorshop?

   F 10 sq ft   H 100 sq ft

   G 200 sq ft   J 350 sq ft

7. When the total costs of V.I.P. Inc. and Floorshop are the same, what is the total cost?

   A $1125.00   C $1950.00

   B $1900.00   D $3187.50

Contractor	Cost to tear out old floor	Cost of new floor per square foot
Smith & Son	$250	$8.00
V.I.P. Inc.	$350	$7.75
Dad's Floors	$150	$8.00
Floorshop	$300	$8.25

8. How many square feet would the Masons need to have installed to make the total cost of Smith & Son the same as the total cost of V.I.P. Inc.?

   F 80 sq ft   H 400 sq ft

   G 100 sq ft   J 1000 sq ft

17

Holt Algebra 1

LESSON 6-2 Reading Strategies
Use a Sequence Chain

Use the sequence chain below to guide you in solving systems of linear equations by the method of substitution.

Step 1: Isolate one variable in one equation.	**Step 2:** Substitute the expression for that variable into the other equation.	**Step 3:** Solve for the variable remaining in that equation.

Sequence Chain: *Solving Systems by Substitution*

Step 4: Substitute the value for that variable into either original equation and solve for the other variable.	**Step 5:** Write the found values for *x* and *y* as an ordered pair (x, y).	**Step 6:** Check your answer by substituting (x, y) into both equations.

Answer each question.

1. Perform Step 1 for $\begin{cases} x + y = 5 \\ -3x + 5 = \frac{1}{2}y \end{cases}$ by solving the first equation for *y*.

____

2. To solve the system $\begin{cases} y = x + 2 \\ 2x + y = 6 \end{cases}$ by substitution, you can start at Step 2. Explain why.

____

3. Why does Step 6 specify substituting into *both* equations?

____

Solve each system of equations by substitution.

4. $\begin{cases} 2x = 3y \\ y = x - 2 \end{cases}$

5. $\begin{cases} x + y = 2 \\ -x = 2y - 7 \end{cases}$

_________________   _________________

Holt Algebra 1

Practice A
Solving Systems by Elimination

Fill in the blanks to solve each system by elimination.

1. $\begin{cases} x + 3y = 14 \\ 2x - 3y = -8 \end{cases}$

Add the equations:
$$x + 3y = 14$$
$$+2x - 3y = 28$$

$3x + \underline{\quad} = 6$

$\underline{\quad} = 6$

$\div \underline{\quad} \quad \div \underline{\quad}$

$x = \underline{\quad}$

Substitute _____ for x in one of the equations:

$x + 3y = 14$

$\underline{\quad} + 3y = 14$

$- \underline{\quad} \quad - \underline{\quad}$

$3y = \underline{\quad}$

$\div 3 \quad \div 3$

$y = \underline{\quad}$

Solution: $(\underline{\quad}, \underline{\quad})$

2. $\begin{cases} 2x + 2y = 4 \\ 3x + 2y = 7 \end{cases}$

Subtract the equations:
$$2x + 2y = 4$$
$$-(3x + 2y = 7)$$

or

$2x + 2y = 4$

$-3x - \underline{\quad} = \underline{\quad}$

$-x + \underline{\quad} = \underline{\quad}$

$-x = \underline{\quad}$

$\div \underline{\quad} \quad \div \underline{\quad}$

$x = \underline{\quad}$

Substitute _____ for x in one of the equations:

$3x + 2y = 7$

$3(\underline{\quad}) + 2y = 7$

$\underline{\quad} + 2y = 7$

$- \underline{\quad} \quad - \underline{\quad}$

$2y = \underline{\quad}$

$\div \underline{\quad} \quad \div \underline{\quad}$

$y = \underline{\quad}$

Solution: $(\underline{\quad}, \underline{\quad})$

3. $\begin{cases} 3x + 4y = 26 \\ x - 2y = -8 \end{cases}$

Multiply the second equation by 2. Then, add the equations:

$\begin{cases} 3x + 4y = 26 \\ 2(x - 2y = -8) \end{cases}$

$3x + 4y = 26$

$+ \underline{\quad} x - \underline{\quad} y = \underline{\quad}$

$\underline{\quad} x + 0 = \underline{\quad}$

$\underline{\quad} x = \underline{\quad}$

$\div \underline{\quad} \quad \div \underline{\quad}$

$x = \underline{\quad}$

Substitute _____ for x in one of the equations:

$x - 2y = -8$

$\underline{\quad} - 2y = -8$

$- \underline{\quad} \quad - \underline{\quad}$

$-2y = \underline{\quad}$

$\div \underline{\quad} \quad \div \underline{\quad}$

$y = \underline{\quad}$

Solution: $(\underline{\quad}, \underline{\quad})$

Solve each system by elimination.

4. $\begin{cases} 3x - 2y = 1 \\ 2x + 2y = 14 \end{cases}$

____

5. $\begin{cases} x + y = 4 \\ 3x + y = 16 \end{cases}$

____

6. $\begin{cases} 3x + 2y = -26 \\ 2x - 6y = -10 \end{cases}$

____

7. The sum of two numbers is -1. When twice the first number and four times the second number are added, the sum is -10. What are the two numbers?

____

Holt Algebra 1

<table><tr><td>**LESSON**
6-3</td><td></td></tr></table>

Practice B
Solving Systems by Elimination

Follow the steps to solve each system by elimination.

1. $\begin{cases} 2x - 3y = 14 \\ 2x + y = -10 \end{cases}$

Subtract the second equation:

$\begin{array}{r} 2x - 3y = 14 \\ -(2x + y = -10) \\ \hline \end{array}$

____

Solve the resulting equation:

$y =$ _______________

Use your answer to find the value of x:

$x =$ _______________

Solution: (_______ , _______)

2. $\begin{cases} 3x + y = 17 \\ 4x + 2y = 20 \end{cases}$

Multiply the first equation by -2. Then, add the equations:

$\begin{array}{r} \underline{}x - \underline{}y = \underline{} \\ + 4x + 2y = 20 \\ \hline \end{array}$

Solve the resulting equation:

$x =$ _______________

Use your answer to find the value of y:

$y =$ _______________

Solution: (_______ , _______)

Solve each system by elimination. Check your answer.

3. $\begin{cases} x + 3y = -7 \\ -x + 2y = -8 \end{cases}$

____

4. $\begin{cases} 3x + y = -26 \\ 2x - y = -19 \end{cases}$

____

5. $\begin{cases} x + 3y = -14 \\ 2x - 4y = 32 \end{cases}$

____

6. $\begin{cases} 4x - y = -5 \\ -2x + 3y = 10 \end{cases}$

____

7. $\begin{cases} y - 3x = 11 \\ 2y - x = 2 \end{cases}$

____

8. $\begin{cases} -10x + y = 0 \\ 5x + 3y = -7 \end{cases}$

____

Solve.

9. Brianna's family spent $134 on 2 adult tickets and 3 youth tickets at an amusement park. Max's family spent $146 on 3 adult tickets and 2 youth tickets. What is the price of a youth ticket? _______________

10. Carl bought 19 apples of 2 different varieties to make a pie. The total cost of the apples was $5.10. Granny Smith apples cost $0.25 each and Gala apples cost $0.30 each. How many of each type of apple did Carl buy? _______________

____

Holt Algebra 1

LESSON 6-3

Practice C
Solving Systems by Elimination

Solve each system by elimination.

1. $\begin{cases} x + y = 2 \\ 2x - y = 7 \end{cases}$

2. $\begin{cases} 3x - 2y = -2 \\ 3x + y = 10 \end{cases}$

3. $\begin{cases} x + y = -7 \\ x - y = 5 \end{cases}$

____

4. $\begin{cases} -3x - 4y = -2 \\ 6x + 4y = 3 \end{cases}$

5. $\begin{cases} 2x - 2y = 14 \\ x + 4y = -13 \end{cases}$

6. $\begin{cases} y - x = 17 \\ 2y + 3x = -11 \end{cases}$

____

7. $\begin{cases} x + 6y = 1 \\ 2x - 3y = 32 \end{cases}$

8. $\begin{cases} -\frac{1}{2}x + y = 4 \\ \frac{1}{3}x - y = -3 \end{cases}$

9. $\begin{cases} 3x + y = -15 \\ 2x - 3y = 23 \end{cases}$

____

10. $\begin{cases} 5x - 2y = -48 \\ 2x + 3y = -23 \end{cases}$

11. $\begin{cases} 4x - 3y = -9 \\ 5x - y = 8 \end{cases}$

12. $\begin{cases} 3x - 3y = -1 \\ 12x - 2y = 16 \end{cases}$

____

13. At a bakery, Riley bought 3 bagels and 2 muffins for $7.25. Karen bought 5 bagels and 4 muffins for $13.25. What is the cost of each item?

____

14. A chemist has a beaker of a 3% acid solution and a beaker of a 7% acid solution. He needs to make 75 mL of a 4% acid solution.

a. Complete the table.

	3% solution	+	7% solution	=	4% solution
Amount of Solution (mL)	x	+	y	=	_______
Amount of Acid (mL)	____x	+	____y	=	0.04(75)

b. Use the information in the table to write a system of linear equations.

____

c. Solve the system of equations to find how much he will use from each beaker.

____

Holt Algebra 1

Reteach
Solving Systems by Elimination

Elimination can be used to solve a system of equations by adding terms vertically. This will cause one of the variables to be eliminated. It may be necessary to multiply one or both equations by some number to use this method.

I. Elimination may require no change to either equation.

$$\begin{cases} 3x + y = 6 \\ 5x - y = 10 \end{cases}$$

Adding vertically will eliminate y.

$$\begin{array}{r} 3x + y = 6 \\ 5x - y = 10 \\ \hline 8x + 0 = 16 \end{array}$$

II. Elimination may require multiplying one equation by an appropriate number.

$$\begin{cases} 2x + 5y = 9 \\ x - 3y = 10 \end{cases}$$

Multiply bottom equation by -2.

$$\begin{array}{r} 2x + 5y = 9 \\ -2(x - 3y) = -2(10) \end{array}$$

$$\begin{array}{r} 2x + 5y = 9 \\ -2x + 6y = -20 \\ \hline 0 + 11y = -11 \end{array}$$

III. Elimination may require multiplying both equations by different numbers.

$$\begin{cases} 5x + 3y = 2 \\ 4x + 2y = 10 \end{cases}$$

Multiply the top by -2 and the bottom by 3.

$$\begin{array}{r} -2(5x + 3y = 2) \\ 3(4x + 2y = 10) \end{array}$$

$$\begin{array}{r} -10x + -6y = -4 \\ 12x + 6y = 30 \\ \hline 2x + 0 = 26 \end{array}$$

Solve each system by elimination.

1. $\begin{cases} 2x - y = 20 \\ 3x + 2y = -19 \end{cases}$

2. $\begin{cases} 3x + 2y = 10 \\ 3x - 2y = 14 \end{cases}$

3. $\begin{cases} x + y = 12 \\ 2x + y = 6 \end{cases}$

4. $\begin{cases} 3x - y = 2 \\ -8x + 2y = 4 \end{cases}$

Holt Algebra 1

Reteach
Solving Systems by Elimination (continued)

A system of equations can be solved by graphing, substitution, or elimination.

- Use graphing if both equations are solved for y, or if you want an estimate of the solution.

- Use substitution if either equation is solved for a variable, or has a variable with a coefficient of 1 or -1.

- Use elimination if both equations have the same variable with the same or opposite coefficients.

It may be necessary to manipulate your equations to get them in any of the three forms above.

Solve $\begin{cases} y = 3 - x \\ 2x - y = 6 \end{cases}$.

> One equation is solved for a variable. Use substitution.

$$2x - y = 6$$
$$2x - (3 - x) = 6$$
$$3x - 3 = 6$$

Substitute x + 2 for y.

$$\underline{+3 \quad +3}$$
$$3x = 9$$
$$x = 3$$

Substitute $x = 3$ into one of the original equations to find the value of y.

$$y = 3 - x$$
$$y = 3 - 3$$
$$y = 0 \qquad \text{The solution is } (3, 0).$$

Solve $\begin{cases} -2x - y = -5 \\ 3x + y = -1 \end{cases}$.

> The equations have the same variable with opposite coefficients. Use elimination.

$$-2x - y = -5$$
$$\underline{3x + y = -1}$$
$$x + 0 = -6$$
$$x = -6$$

Substitute $x = -6$ into one of the original equations to find the value of y.

$$3x + y = -1$$
$$3(-6) + y = -1$$
$$-18 + y = -1$$
$$\underline{+18 \qquad +18}$$
$$y = 17$$

The solution is $(-6, 17)$.

Solve each system by any method.

5. $\begin{cases} y = x + 3 \\ -2x + y = -4 \end{cases}$

6. $\begin{cases} 4x + y = 10 \\ -2x - y = 4 \end{cases}$

7. $\begin{cases} 2x + y = 8 \\ 3x + 5y = 5 \end{cases}$

___________   ___________   ___________

Holt Algebra 1

Challenge
Solving Systems by Elimination

The elimination method can also be used for a system of three equations in three unknowns.

Three camp leaders purchased equipment for a camping trip.
Max bought 10 sleeping bags, 2 tents, and 1 can of bug repellant for $885.
Carlos bought 5 sleeping bags, 4 tents, and 1 can of bug repellant for $865.
Amy bought 9 sleeping bags, 6 tents, and 6 cans of bug repellant for $1410.
If they made their purchases at the same store, how much did each item cost?

1. Write the 3 equations:
$$10x + 2y + z = 885$$
____

____

2. Subtract the second equation from the first. _______________________

3. Multiply the second equation by -6
   and add the second and third equations. _______________________

4. The equations in steps 2 and 3 form a linear
   system in two variables. Solve this system for x. _______________________

5. Substitute the value of x into the first two
   equations. Write the resulting system. _______________________

6. Solve the system in problem 5 for y and z. _______________________

7. Write the cost of each item. _______________________

Holt Algebra 1

Problem Solving

LESSON 6-3

Solving Systems by Elimination

Write the correct answer.

1. Mr. Nguyen bought a package of 3 chicken legs and a package of 7 chicken wings. Ms. Dawes bought a package of 3 chicken legs and a package of 6 chicken wings. Mr. Nguyen bought 45 ounces of chicken. Ms. Dawes bought 42 ounces of chicken. How much did each chicken leg and each chicken wing weigh?

____

____

2. Jayce bought 2 bath towels and returned 3 hand towels. His sister Jayna bought 3 bath towels and 3 hand towels. Jayce's bill was $5. Jayna's bill was $45. What are the prices of a bath towel and a hand towel?

____

____

3. The Lees spent $31 on movie tickets for 2 adults and 3 children. The Macias spent $26 on movie tickets for 2 adults and 2 children. What are the prices for adult and child movie tickets?

____

____

4. Last month Stephanie spent $57 on 4 allergy shots and 1 office visit. This month she spent $9 after 1 office visit and a refund for 2 allergy shots from her insurance company. How much does an office visit cost? an allergy shot?

____

____

Use the chart below to answer questions 5–6. Select the best answer. The chart shows the price per pound for dried fruit.

Dried Fruit Price List			
Pineapple	Apple	Mango	Papaya
$7.50/lb	$7.00/lb	$8.00/lb	$7.25/lb

5. A customer bought 5 pounds of mango and papaya for $37.75. How many pounds of each fruit did the customer buy?

   A 2 lbs mango and 3 lbs papaya

   B 3 lbs mango and 2 lbs papaya

   C 1 lb mango and 4 lbs papaya

   D 4 lbs mango and 1 lb papaya

6. A store employee made two gift baskets of dried fruit, each costing $100. The first basket had 12 pounds of fruit x and 2 pounds of fruit y. The second basket had 4 pounds of fruit x and 9 pounds of fruit y. Which two fruits did the employee use in the baskets?

   F pineapple and apple

   G apple and mango

   H mango and papaya

   J papaya and pineapple

Holt Algebra 1

Reading Strategies
Connecting Concepts

When solving systems of linear equations using elimination, you will sometimes need to multiply one or both equations by a factor in order to get the same coefficients for a variable. This process is very similar to getting a common denominator for fractions. Look at the example below.

$\begin{cases} 6x - 5y = 16 \\ 4x - 3y = 12 \end{cases}$ ⟹ To eliminate the *x*-terms, you need to get the same or opposite coefficients for *x* in both equations.

Add: $\dfrac{5}{6} + \dfrac{1}{4} = ?$ ⟹ Think about finding a common denominator for 4 and 6.

4: 4, 8, 12, 16, 20
6: 6, 12, 18, 24, 30 ⟹ Find the least common multiple (LCM) of 4 and 6 by listing their multiples in order. The LCM is 12.

$\begin{cases} 2(6x - 5y = 16) \\ 3(4x - 3y = 12) \end{cases}$ ⟹ Determine what you have to multiple 4 and 6 by to get 12. Multiply each equation by the appropriate number.

$\begin{cases} 12x - 10y = 32 \\ 12x - 9y = 36 \end{cases}$ ⟹ Now either add or subtract the equations. In this case, you will subtract the equations.

1. Describe how you would get common *y*-coefficients (instead of *x*-) in the example above.

____

____

2. Show how to get a set of common *y*-coefficients for the system $\begin{cases} 9x - 10y = 7 \\ 5x + 8y = 31 \end{cases}$.

____

Solve each system of equations by elimination.

3. $\begin{cases} 9x - 2y = 15 \\ 4x + 3y = -5 \end{cases}$

4. $\begin{cases} 2x - 3y = 50 \\ 7x + 8y = -10 \end{cases}$

____

Holt Algebra 1

LESSON 6-4
Practice A
Solving Special Systems

Solve each system of linear equations. Tell whether the system has no solution or infinitely many solutions.

1. $\begin{cases} 2x + y = 1 \\ 2x + y = -3 \end{cases}$

2. $\begin{cases} y = 5x + 2 \\ y - 5x = 2 \end{cases}$

____

3. $\begin{cases} y - 3x + 2 = 0 \\ 2 = -y + 3x \end{cases}$

4. $\begin{cases} x + y = 4 \\ y - 4 = 1 - x \end{cases}$

____

Give the number of solutions to each system. Then classify the system as "consistent, independent", "consistent, dependent", or "inconsistent".

5. $\begin{cases} y = 2(x + 1) \\ y - 2x = 2 \end{cases}$

6. $\begin{cases} y - 4x + 5 = 0 \\ 4x = y - 1 \end{cases}$

____

7. Marquis opens a savings account with $60 and adds $20 each month. His brother Jibran adds $20 each month to the savings account that his grandmother opened with $60. If the brothers continue to make deposits to their savings accounts at the same rate, when will they have the same amount of money? Explain.

____

____

Holt Algebra 1

Practice B
Solving Special Systems

Solve each system of linear equations.

1. $\begin{cases} y = 2x - 3 \\ y - 2x = -3 \end{cases}$

2. $\begin{cases} 3x + y = 4 \\ -3x = y - 7 \end{cases}$

____

3. $\begin{cases} y = -4x + 1 \\ 4x = -y - 6 \end{cases}$

4. $\begin{cases} y - x + 3 = 0 \\ x = y + 3 \end{cases}$

____

Classify each system. Give the number of solutions.

5. $\begin{cases} y = 3(x - 1) \\ -y + 3x = 3 \end{cases}$

6. $\begin{cases} y - 2x = 5 \\ x = y - 3 \end{cases}$

____

7. Sabina and Lou are reading the same book. Sabina reads 12 pages a day. She had read 36 pages when Lou started the book, and Lou reads at a pace of 15 pages per day. If their reading rates continue, will Sabina and Lou ever be reading the same page on the same day? Explain.

8. Brandon started jogging at 4 miles per hour. After he jogged 1 mile, his friend Anton started jogging along the same path at a pace of 4 miles per hour. If they continue to jog at the same rate, will Anton ever catch up with Brandon? Explain.

Holt Algebra 1

LESSON 6-4
Practice C
Solving Special Systems

Solve each system of linear equations.

1. $\begin{cases} y + 2x + 4 = 0 \\ 2x = -y - 4 \end{cases}$

2. $\begin{cases} 5x + y = 8 \\ -5x = 8 + y \end{cases}$

3. $\begin{cases} 2x - y = 4 \\ 1 = y - 2x + 5 \end{cases}$

4. $\begin{cases} y = -x - 6 \\ y - 2x = -3x + 6 \end{cases}$

Classify each system. Give the number of solutions.

5. $\begin{cases} y + 2(x - 3) = 0 \\ 2x = -y - 3 \end{cases}$

6. $\begin{cases} y + 3x = -1 \\ x = y + 3x - 1 \end{cases}$

7. At a factory, Jin assembles 12 parts each minute. He has assembled 156 parts when Summer starts on the line, assembling at a pace of 15 parts per minute. If their assembly rates continue, will Summer ever catch up to Jin? Explain.

8. Kat is comparing monthly sales at her bookstore with those of her competitor, Gill. If the sales rates continue, will Kat's book sales ever catch up with her competitor's? Explain.

Kat's sales	112	118	124	130
Gill's sales	138	144	150	156

Holt Algebra 1

Reteach

LESSON 6-4

Solving Special Systems

When solving equations in one variable, it is possible to have one solution, no solutions, or infinitely many solutions. The same results can occur when graphing systems of equations.

Solve $\begin{cases} 4x + 2y = 2 \\ 2x + y = 4 \end{cases}$.

Multiplying the second equation by -2 will eliminate the x-terms.

$$4x + 2y = 2$$
$$-2(2x + y = 4) \longrightarrow$$

$$4x + 2y = 2$$
$$\underline{-4x - 2y = -8}$$
$$0 + 0 = -6$$
$$0 = -6 \ ✗$$

The equation is a contradiction. **There is no solution.**

Solve $\begin{cases} y = 4 - 3x \\ 3x + y = 4 \end{cases}$.

Because the first equation is solved for a variable, use substitution.

$$3x + y = 4$$
$$3x + (4 - 3x) = 4 \quad \textit{Substitute } 4 - 3x \textit{ for } y$$
$$0 + 4 = 4$$
$$4 = 4 \ ✓$$

The equation is true for all values of x and y. **There are infinitely many solutions.**

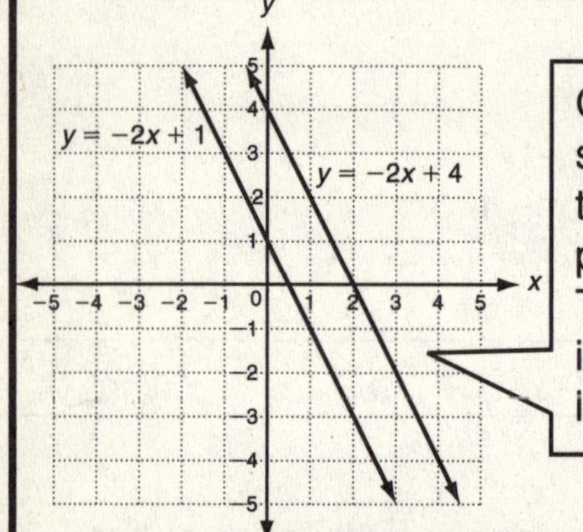

Graphing the system shows that these are parallel lines. They will never intersect, so there is no solution.

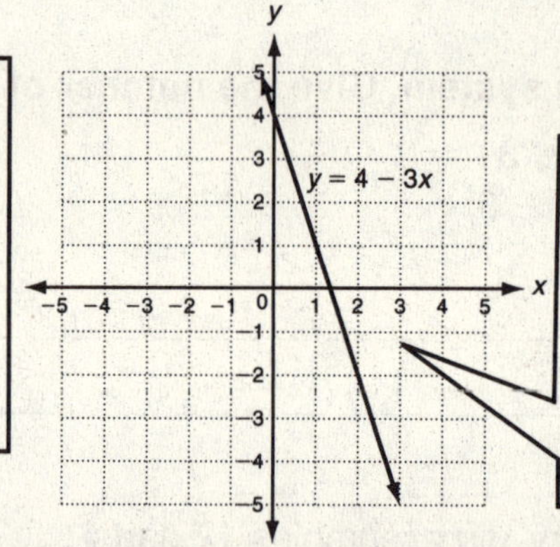

The slopes and y-intercepts are the same. These are the same line.

Solve each system of linear equations algebraically.

1. $\begin{cases} y = 3x \\ 2y = 6x \end{cases}$

2. $\begin{cases} y = 2x + 5 \\ y - 2x = 1 \end{cases}$

3. $\begin{cases} 3x - 2y = 9 \\ -6x + 4y = 1 \end{cases}$

_____________________ _____________________ _____________________

Holt Algebra 1

LESSON 6-4 Reteach
Solving Special Systems (continued)

A system of linear equations can be classified in three ways.

I. Consistent and independent one solution different slopes	Example: $$\begin{cases} y = x + 3 \\ y = -x + 6 \end{cases}$$	
II. Consistent and dependent infinitely many solutions same slope, same y-intercepts	Example $$\begin{cases} y = 3x + 4 \\ y - 3x = 4 \end{cases}$$	
III. Inconsistent no solutions same slope, different y-intercepts	Example $$\begin{cases} y = 2x + 5 \\ y = 2x + 2 \end{cases}$$	

Classify each system below by comparing the slopes and y-intercepts. Then give the number of solutions.

4. $\begin{cases} y = -3x - 2 \\ y = -3x - 4 \end{cases}$

5. $\begin{cases} y = 2x + 5 \\ y = 5 + 2x \end{cases}$

6. $\begin{cases} y = -4x + 3 \\ y = 2x + 7 \end{cases}$

_______________ _______________ _______________

_______________ _______________ _______________

Classify each system and give the number of solutions. If there is one solution, provide it.

7. $\begin{cases} y = 2x + 8 \\ y - 4x = 8 \end{cases}$

8. $\begin{cases} y + 3x - 2 = 0 \\ 9x + 3y = 6 \end{cases}$

_______________ _______________

_______________ _______________

Holt Algebra 1

LESSON 6-4 Challenge
Pick a Path

The puzzle below contains 23 linear equations. You goal is to find paths
through the puzzle such that each pair of equations along your path forms
a special type of system.

You must begin in the **Start** square, and end in the **Finish** square. Moves
can be horizontal, vertical, or diagonal between adjacent squares. Each
individual path cannot cross itself, but your answers to 1 and 2 might cross.

	1	**2**	**3**	**4**
Start →	$y = x + 2$	$3x = 3y - 6$	$3x + y = -1$	$8y = 8x + 16$
5 $y = 2x + 3$	**6** $y = 3x - 1$	**7** $y - x = 2$	**8** $\frac{1}{3}x + \frac{1}{3}y = 1$	**9** $3x = y + 1$
10 $2y = 6x + 6$	**11** $2x - y = -3$	**12** $2x + 2y = 4$	**13** $2y - 4 = 2x$	**14** $x - y = -2$
15 $\frac{1}{2}y = \frac{3}{2}x + \frac{1}{2}$	**16** $3y = 2x + 1$	**17** $3x - y = 4$	**18** $\frac{1}{6}y = \frac{1}{3}x + \frac{1}{2}$	**19** $\frac{1}{2}y = \frac{1}{2}x + 1$
20 $10x + 15y = 5$	**21** $-12x + 4y = 8$	**22** $\frac{1}{2}y = \frac{1}{3}x + \frac{1}{6}$	**23** $x = \frac{1}{3}y + 1$	**→ Finish**

**Find a path through the puzzle such that each pair of equations
forms...**

1. ...a consistent and dependent system.

____

2. ...an inconsistent system.

____

Holt Algebra 1

Problem Solving
LESSON 6-4
Solving Special Systems

Write the correct answer.

1. Tyra and Charmian are training for a bike race. Tyra has logged 256 miles so far and rides 48 miles per week. Charmian has logged 125 miles so far and rides 48 miles per week. If these rates continue, will Tyra's distance ever equal Charmian's distance? Explain.

2. Metroplexpress and Local Express are courier companies. Metroplexpress charges $15 to pick up a package and $0.50 per mile. Local Express charges $10 to pick up a package and $0.55 per mile. Classify this system and find its solution, if any.

3. The Singhs start savings accounts for their twin boys. The accounts earn 5% annual interest. The initial deposit in each account is $200. Classify this system and find its solution, if any.

4. Frank earns $8 per hour. Madison earns $7.50 per hour. Frank started working after Madison had already earned $300. If these rates continue, will Frank's earnings ever equal Madison's earnings? If so, when?

Select the best answer.

5. A studio apartment at The Oaks costs $400 per month plus a $350 deposit. A studio apartment at Crossroads costs $400 per month plus a $300 deposit. How many solutions does this system have?

   A no solutions

   B 1 solution

   C 2 solutions

   D an infinite number of solutions

6. Jane and Gary are both landscape designers. Jane charges $75 for a consultation plus $25 per hour. Gary charges $50 for a consultation plus $30 per hour. For how many hours will Jane's charges equal Gary's charges?

   F never

   G after 2 hours

   H after 5 hours

   J always

7. A tank filled with 75 liters of water loses 0.5 liter of water per hour. A tank filled with 50 liters of water loses 0.1 liter of water per hour. How would this system be classified?

   A inconsistent

   B dependent

   C consistent and independent

   D consistent and dependent

8. Simon is 3 years older than Renata. Five years ago, Renata was half as old as Simon is now. How old are Simon and Renata now?

   F Simon is 13 and Renata is 10.

   G Simon is 15 and Renata is 10.

   H Simon is 16 and Renata is 8.

   J Simon is 16 and Renata is 13.

Holt Algebra 1

Reading Strategies

LESSON 6-4

Use a Table

The table below can help you answer questions about linear systems.

Classification	Number of Solutions	Similarities and Differences in $y = mx + b$	Description of Graphed Lines	Result of Solving with Algebra
Consistent, Independent	1	different slopes (m)	intersecting	values for x and y ex. $x = 3$, $y = -5$
Consistent, Dependent	infinitely many	same slope (m) same y-int. (b)	coincident	identity statement ex. $4 = 4$
Inconsistent	0	same slope (m) different y-int. (b)	parallel	false statement ex. $-2 = 3$

1. Mary Kate solved a system by elimination as shown below. Classify the system.

$$\begin{cases} 2x + 3y = 5 \\ 2x + 3y = 7 \end{cases} \Rightarrow \begin{matrix} 2x + 3y = 5 \\ -(2x + 3y = 7) \\ \hline 0 = -2 \end{matrix}$$

____

2. Raul solved a system of equations by substitution. He ended up with an identity statement. How many solutions does his system have?

____

3. How many solutions does a system have if, when graphed, the lines are the same line?

____

4. Two equations in a system have the same slope. Which classification can NOT describe the system?

____

5. The graph of a system consists of two intersecting lines. How many solutions does the system have?

____

Classify each system, give the number of solutions, and describe its graph.

6. $\begin{cases} y = -3x \\ y = -3x + 2 \end{cases}$ **7.** $\begin{cases} y = -x + 4 \\ x + y = 4 \end{cases}$ **8.** $\begin{cases} y = 2x - 1 \\ x = y \end{cases}$

_______________   _______________   _______________

_______________   _______________   _______________

_______________   _______________   _______________

    Holt Algebra 1

LESSON 6-5
Practice A
Solving Linear Inequalities

Use substitution to tell whether the ordered pair is a solution of the given inequality.

1. $(3, 4)$; $y > x + 2$ **2.** $(4, 2)$; $y \le 2x - 3$ **3.** $(2, -1)$; $y < -x$

_______________   _______________   _______________

Rewrite each linear inequality in slope-intercept form. Then graph the solutions in the coordinate plane.

4. $y - x \le 3$

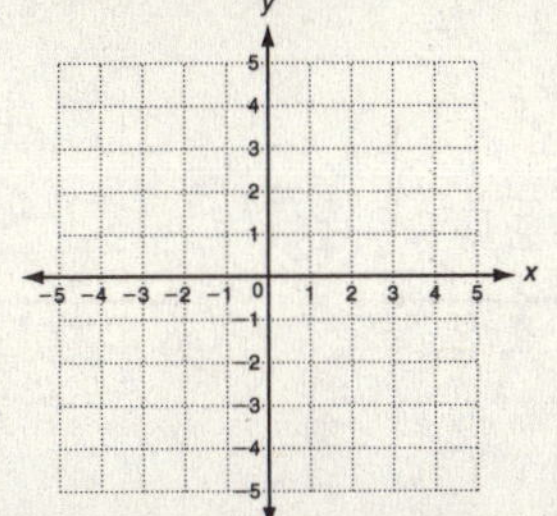

____

5. $6x + 2y > -2$

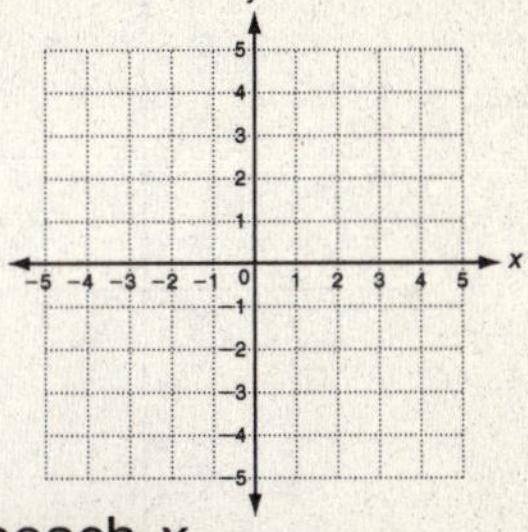

____

6. Trey is choosing yogurt at the grocery store. The flavors he likes are peach x and blueberry y. His mother said he may choose at most 8 containers of yogurt.

   a. Write an inequality to describe the situation.

   ____

   b. Graph the solutions.

   c. Give two possible combinations of peach and blueberry yogurt that Trey can choose.

   ____

Write an inequality to represent each graph.

7.

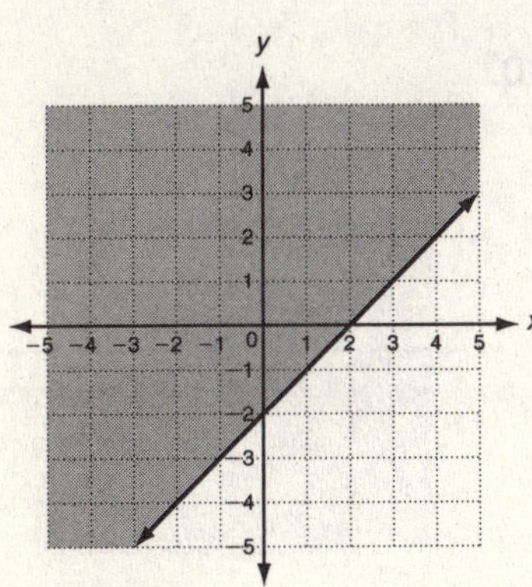

____

8.

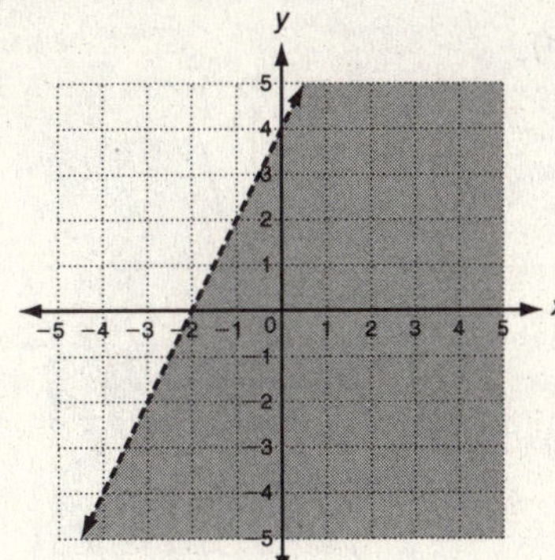

____

9.

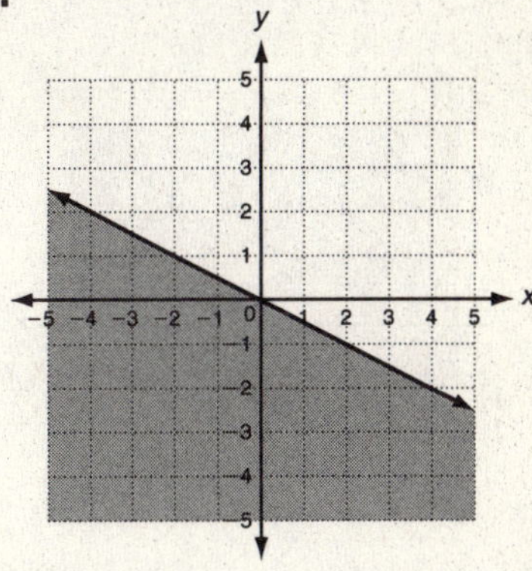

____

       Holt Algebra 1

Practice B
Solving Linear Inequalities

Tell whether the ordered pair is a solution of the given inequality.

1. $(1, 6)$; $y < x + 6$ **2.** $(-3, -12)$; $y \geq 2x - 5$ **3.** $(5, -3)$; $y \leq -x + 2$

____________   ____________   ____________

Graph the solutions of each linear inequality.

4. $y \leq x + 4$ **5.** $2x + y > -2$ **6.** $x + y - 1 < 0$

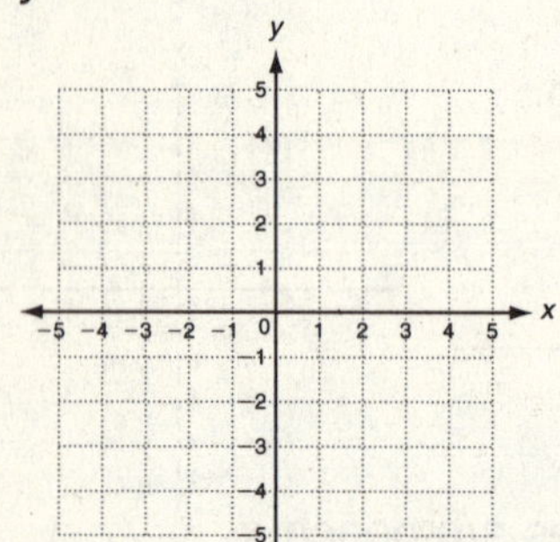 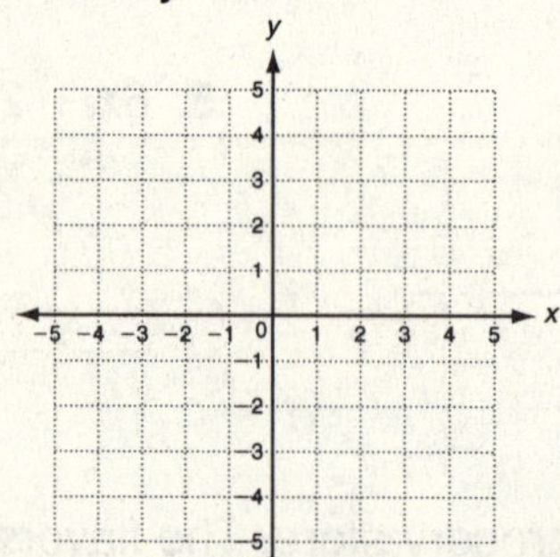 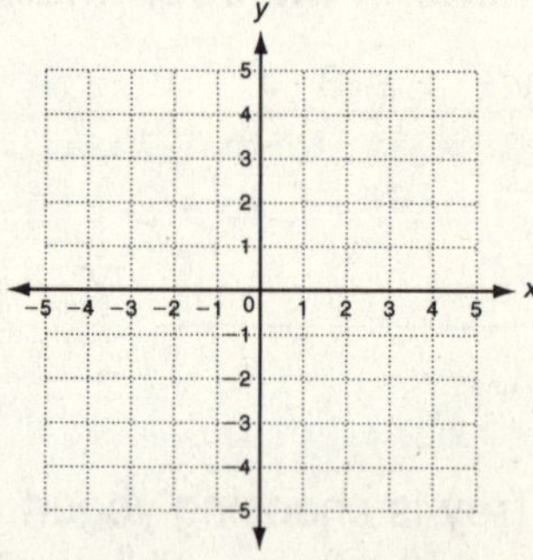

7. Clark is having a party at his house. His father has allowed him to spend at most $20 on snack food. He'd like to buy chips that cost $4 per bag, and pretzels that cost $2 per bag.

a. Write an inequality to describe the situation.

____

b. Graph the solutions.

c. Give two possible combinations of bags of chips and pretzels that Clark can buy.

____

Write an inequality to represent each graph.

8.   **9.**   **10.**

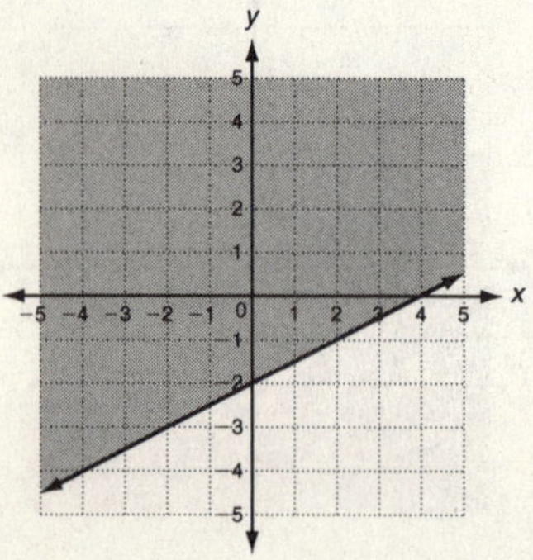 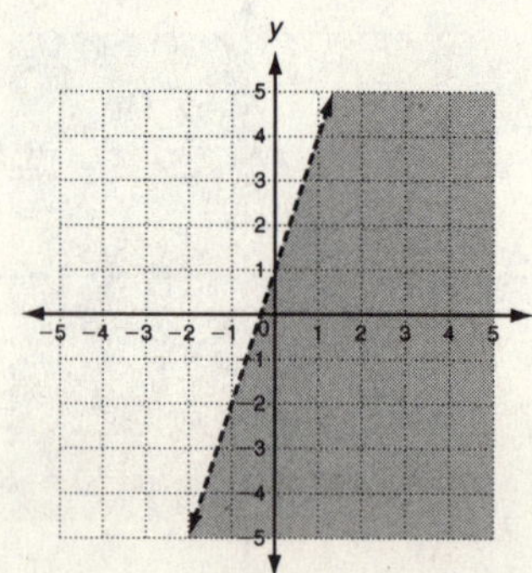 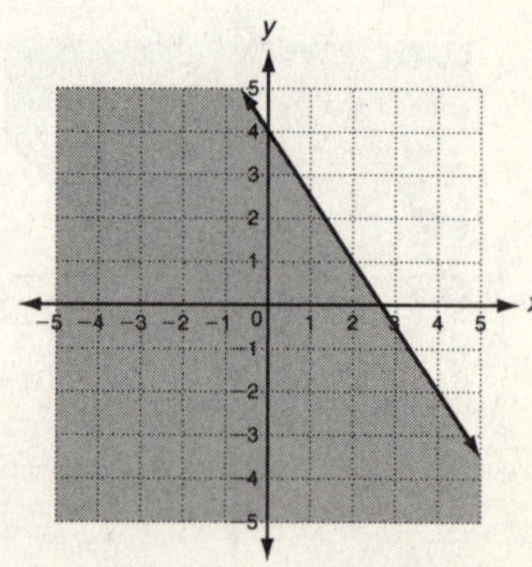

____________   ____________   ____________

Holt Algebra 1

Practice C
Solving Linear Inequalities

Tell whether the ordered pair is a solution of the given inequality.

1. $(-1, -4); y \geq 2x - 1$ **2.** $(-6, 2); y < -x - 4$ **3.** $(4, -8); y \leq \frac{1}{2}x + 5$

_______________ _______________ _______________

Graph the solutions of each linear inequality.

4. $y \leq x + 2$ **5.** $-3x < y$ **6.** $2x - y - 4 > 0$

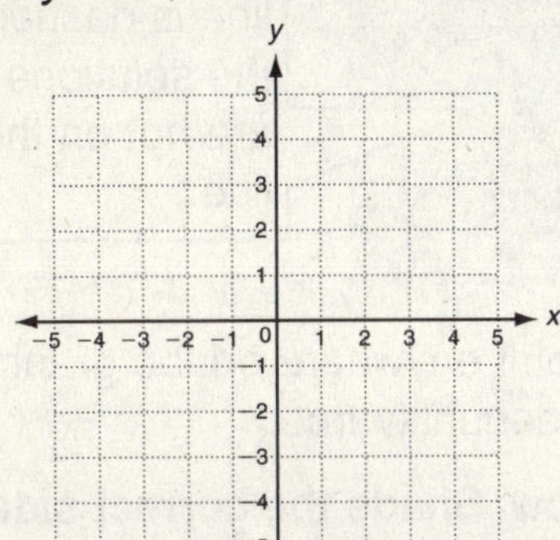 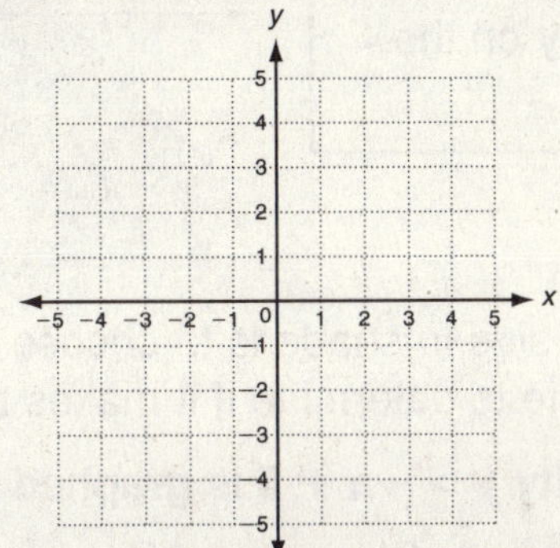 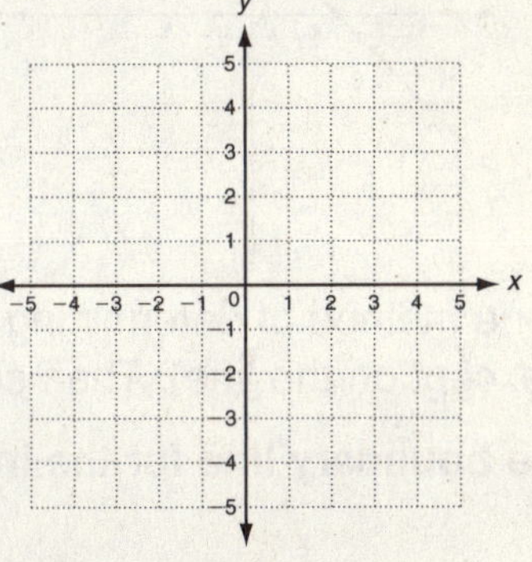

7. Adam is ordering helium balloons for his sister's birthday. He has up to $15 to spend. Decorative balloons cost $3.00 each and solid colored balloons cost $0.50 each.

  a. Write an inequality to describe the situation.

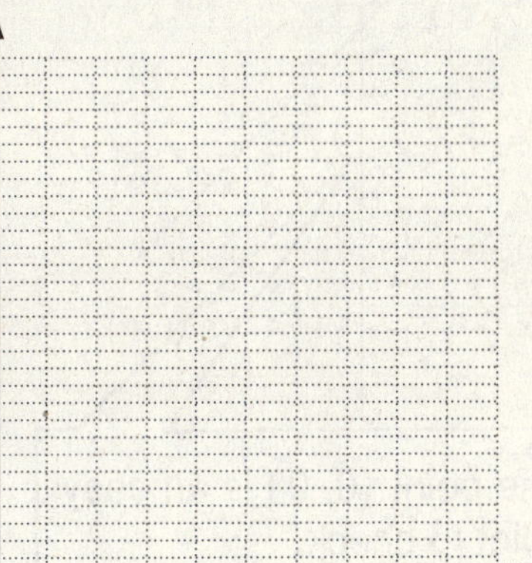

____

  b. Graph the solutions.

  c. Give two possible combinations of decorative
    and solid colored balloons Adam can order.

____

____

Write an inequality to represent each graph.

8.   **9.**   **10.**

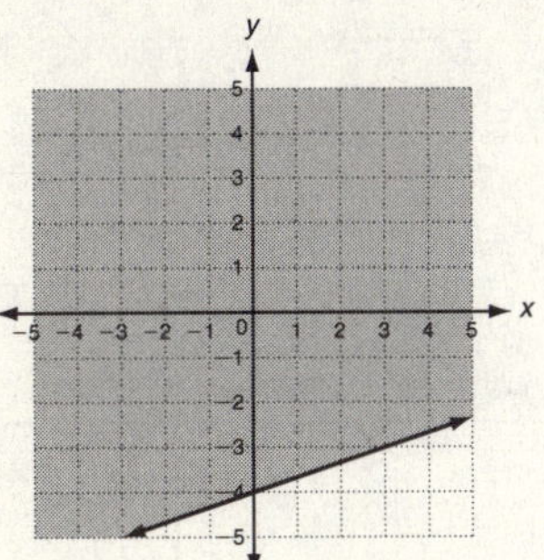 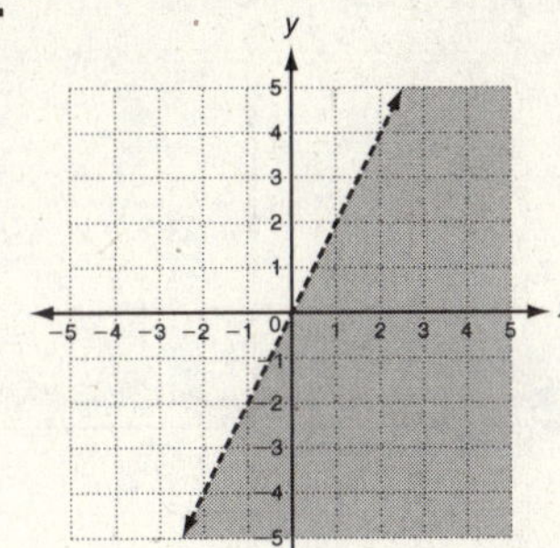 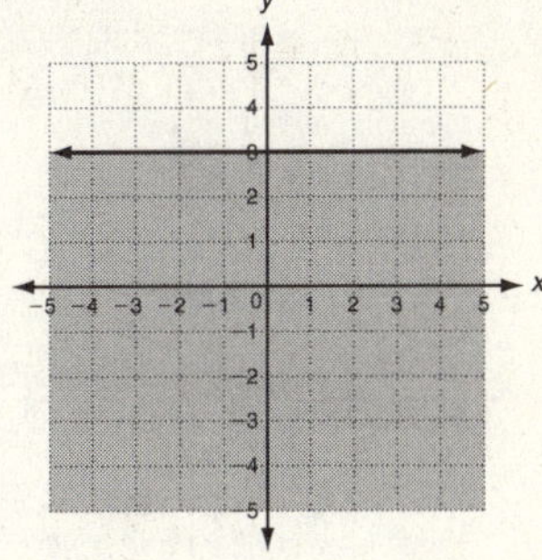

_______________ _______________ _______________

Holt Algebra 1

<table><tr><td>LESSON
6-5</td><td>

Reteach
Solving Linear Inequalities
</td></tr></table>

When graphing an equation, the solutions are all the points on the line. When graphing an inequality, the solutions are all the points above or below the line (and may include the line).

Graph $y = x + 4$.   **Graph $y < x + 4$.**

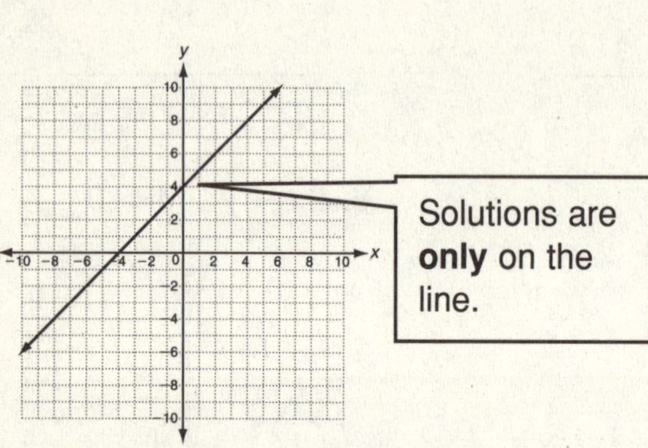

Solutions are **only** on the line.

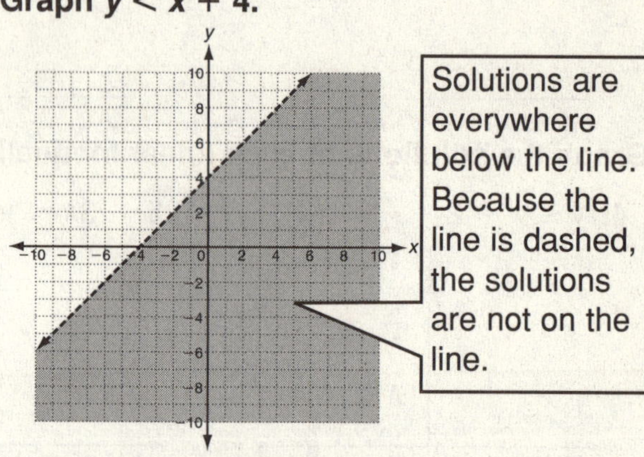

Solutions are everywhere below the line. Because the line is dashed, the solutions are not on the line.

One method of determining which side to shade is to choose a point anywhere on the graph (except on the line). Then substitute to determine if it makes the inequality true.

The boundary line for the inequality $y > -x + 5$ is graphed below. Shade the correct side.

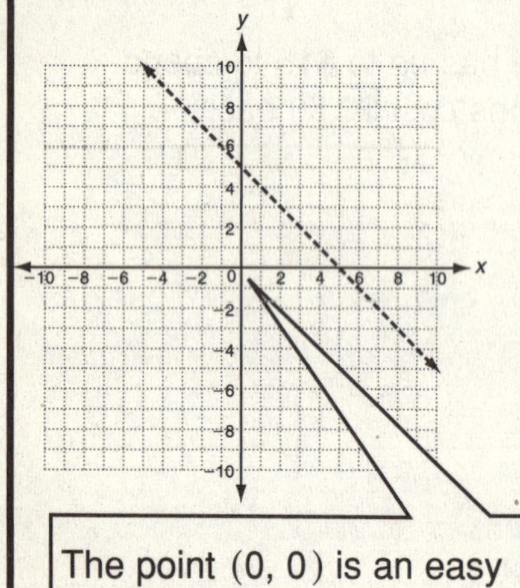

The point $(0, 0)$ is an easy point to check.

Step 1: Choose a point.

Step 2: Substitute $(0, 0)$ in the inequality $y > -x + 5$.

$$y > -x + 5$$
$$0 \overset{?}{>} -0 + 5$$
$$0 \overset{?}{>} 5$$

The statement is false.

Step 3: Because $(0, 0)$, which is below the line, resulted in a false statement, it is not a solution. Shade above the line.

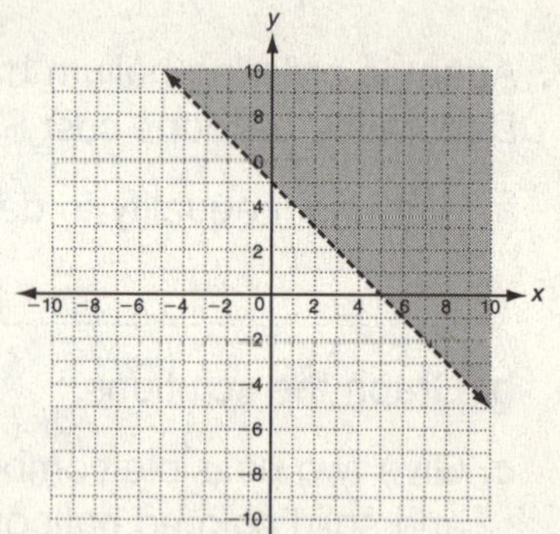

The boundary lines for each inequality are graphed below. Shade the correct side

1. $y > 5x + 7$ **2.** $y < -2x - 9$ **3.** $x > 3$

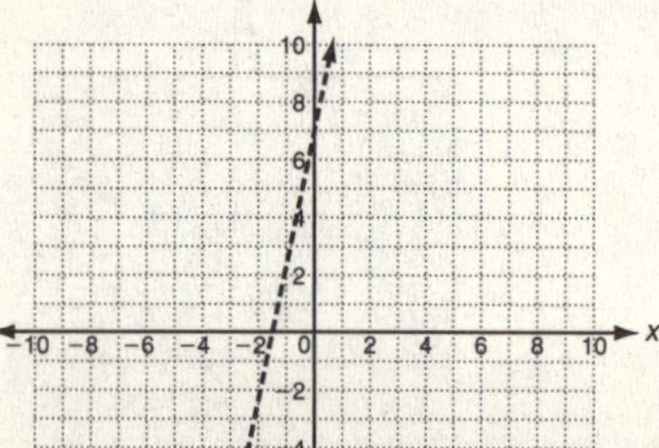

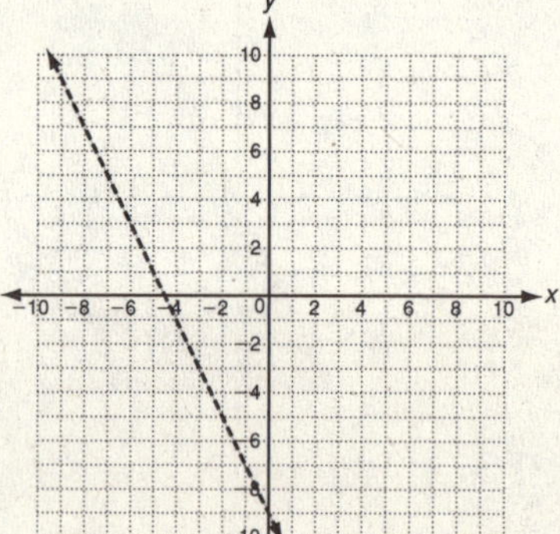

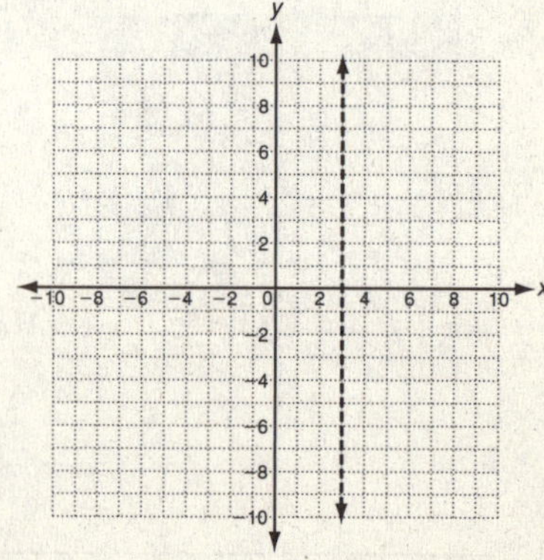

Holt Algebra 1

Reteach

Solving Linear Inequalities (continued)

To graph a linear inequality:

Step 1: Solve the inequality for y.

Step 2: Graph the boundary line. If $\leq$, or $\geq$ use a solid line. If $<$ or $>$ use a dashed line.

Step 3: Determine which side to shade.

Graph the solutions of $2x + y \leq 4$.

Step 1: Solve for y.

$$2x + y \leq 4$$
$$\underline{-2x \qquad -2x}$$
$$y \leq -2x + 4$$

Step 2: Graph the boundary line.

Use a solid line for $\leq$.

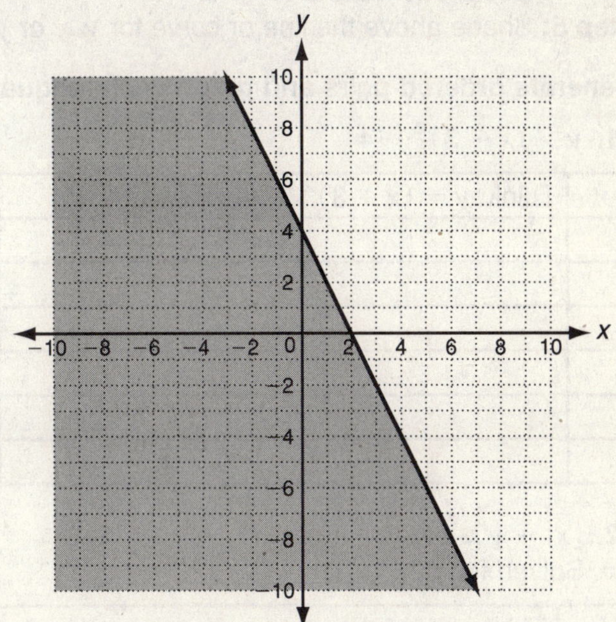

Step 3: Determine which side to shade.

Substitute $(0, 0)$ into $2x + y \leq 4$.

$$2x + y \leq 4$$
$$2(0) + 0 \overset{?}{\leq} 4$$
$$0 \overset{?}{\leq} 4.$$ The statement is true. Shade the side that contains the point $(0, 0)$.

Graph the solutions of each linear inequality.

4. $y - x < 3$

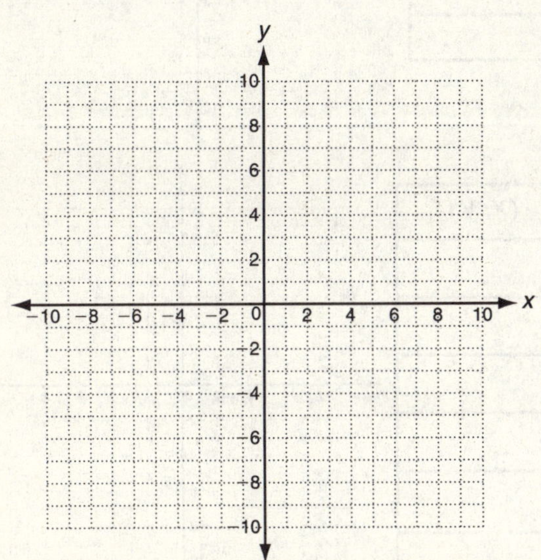

5. $x + y + 2 \geq 0$

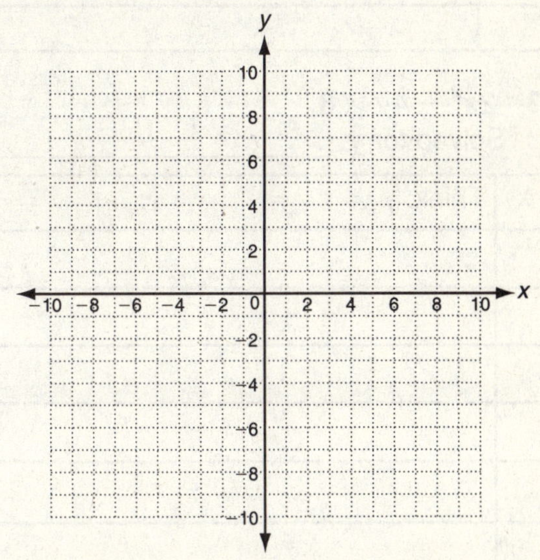

Holt Algebra 1

Challenge
Graphing Non-Linear Inequalities

To graph non-linear inequalities, follow the steps below.

Step 1: Solve the inequality for *y*.

Step 2: Write the related equation and generate ordered pairs.

Step 3: Plot enough points to see a pattern.

Step 4: Connect the points with a line or curve. Use a solid line for $\le$ or $\ge$.
Use a dashed line for $<$ or $>$.

Step 5: Shade above the line or curve for $y >$ or $y \ge$. Shade below for $y <$ or $y \le$.

Generate ordered pairs and graph each inequality.

1. $y > (x + 3)^2 - 4$

x	Think: $y = (x + 3)^2 - 4$	(*x*, *y*)

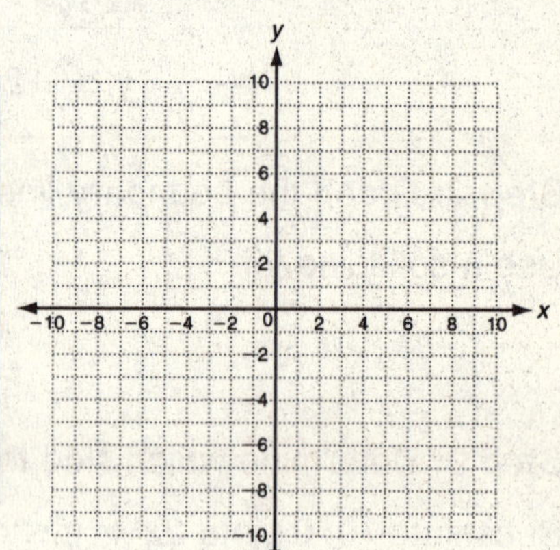

2. $|x| + y \le 5$
Solved for *y*: _____________________

x	Think: $y =$	(*x*, *y*)

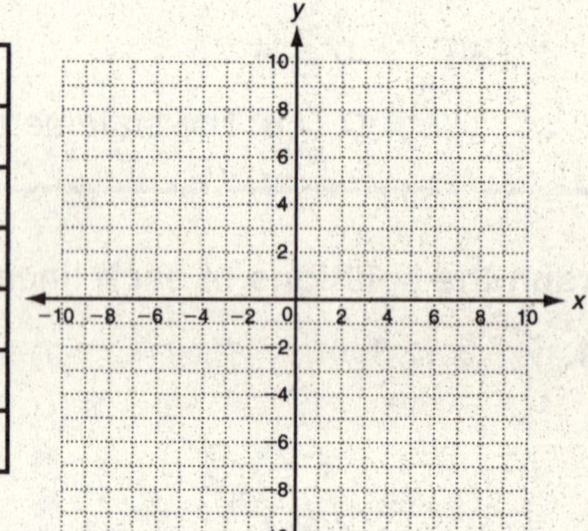

3. $2x^2 - 4y > 8$
Solved for *y*: _____________________

x	Think: $y =$	(*x*, *y*)

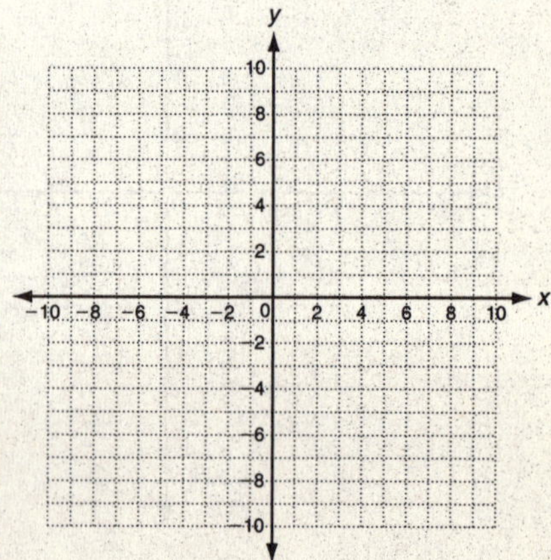

Holt Algebra 1

Problem Solving

LESSON 6-5

Solving Linear Inequalities

Write the correct answer.

1. Shania would like to give $5 gift cards and $4 teddy bears as party favors. Sixteen people have been invited to the party. Shania has $100 to spend on party favors. Write and graph an inequality to find the number of gift cards x and teddy bears y Shania could purchase.

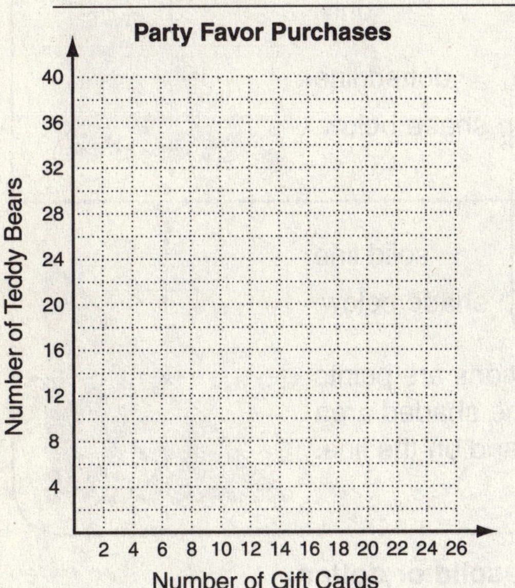

2. Hank has 20 yards of lumber that he can use to build a raised garden. Write and graph a linear inequality that describes the possible lengths and widths of the garden. If Hank wants the dimensions to be whole numbers only, what dimensions would produce the largest area?

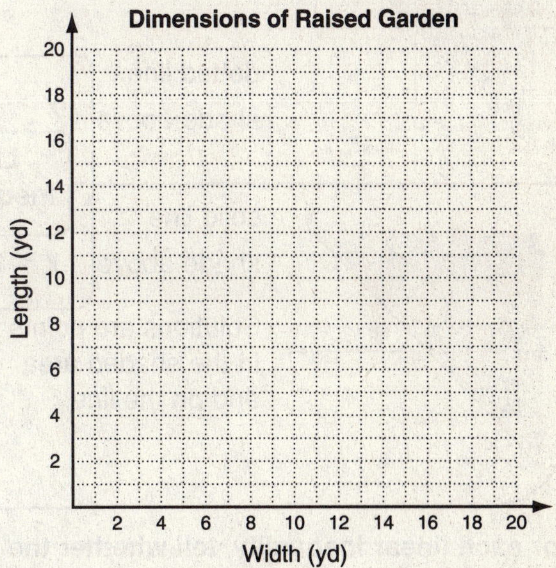

Select the best answer.

3. The royalties for the high school play are $250. Tickets to the play cost $5 for students and $8 for nonstudents. What linear inequality describes the number of student and nonstudent tickets that need to be sold so that the drama class can pay the royalties?

  A $5x + 8y \geq 250$ **C** $5xy + 8 < 250$

  B $5x + 8y > 250$ **D** $5xy + 8 \geq 250$

4. The inequality $x + y \leq 8$ describes the amounts of two juices Annette combines to make a smoothie. Which is a solution to the inequality?

  F $(3, 6)$ **H** $(7, 2)$

  G $(6, 1)$ **J** $(0, 10)$

5. A baker is making chocolate and lemon pound cakes. He can make at most 12 cakes at one time. Which inequality describes the situation?

  A $x + y > 12$ **C** $x + y \leq 12$

  B $x + y \geq 12$ **D** $x + y < 12$

6. Erasmus is the master gardener for a university. He wants to plant a mixture of purple and yellow pansies at the west entrance to the campus. From past experience, Erasmus knows that fewer than 350 pansies will fit in the planting area. Which inequality describes the situation?

  F $x + y \geq 350$ **H** $x + y \leq 350$

  G $x + y > 350$ **J** $x + y < 350$

Holt Algebra 1

Reading Strategies
LESSON 6-5
Use Graphic Aids

There are infinitely many solutions for linear inequalities. That's why the solutions are shown as a graph on a coordinate plane. There are four possible ways to draw the line and shade the correct half plane. These correspond to the four inequality symbols $>$, $\geq$, $<$, $\leq$.
Use the graphic aid below as a guide.

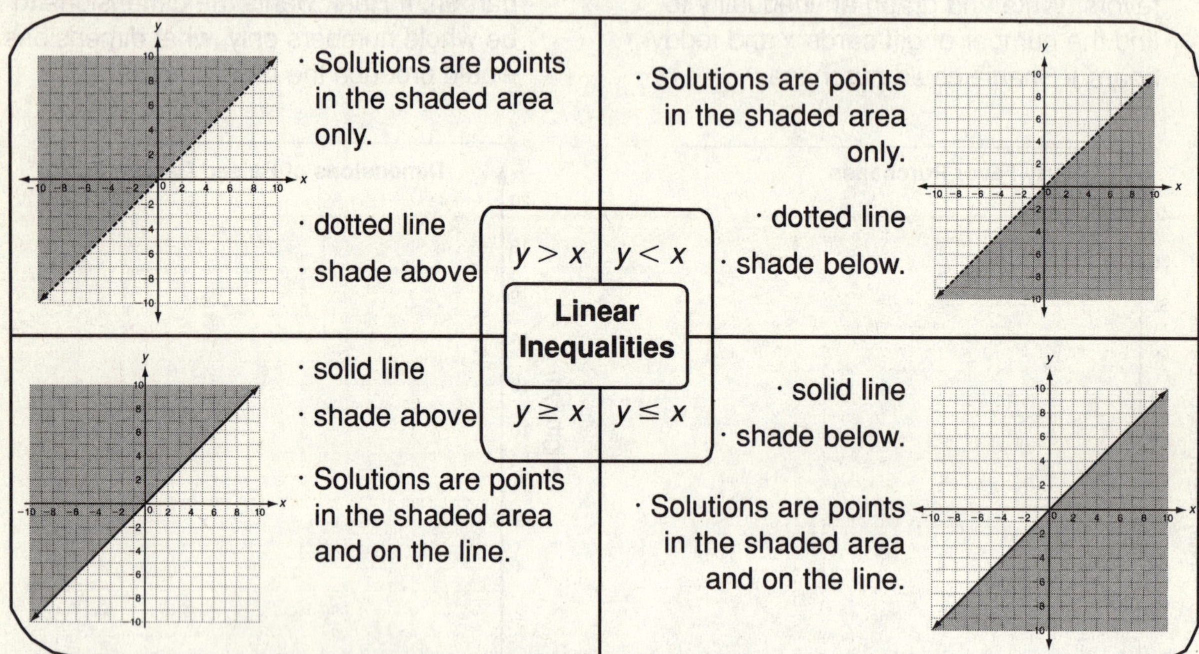

For each linear inequality, tell whether the graph is a solid or dotted line, and whether the shading is above or below the line.

1. $y \leq x + 7$ **2.** $y > -5x$ **3.** $y \geq 3x - 2$

_________________   _________________   _________________

_________________   _________________   _________________

Graph each linear inequality. Name one point that is a solution, and one point that is not a solution.

4. $y \geq -\frac{1}{2}x + 3$
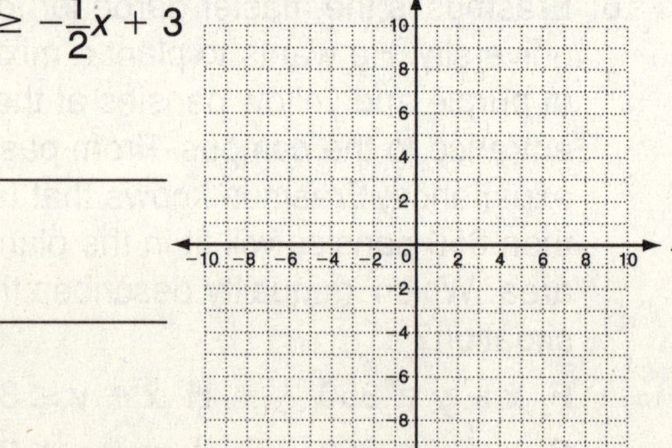

5. $y < x + 5$
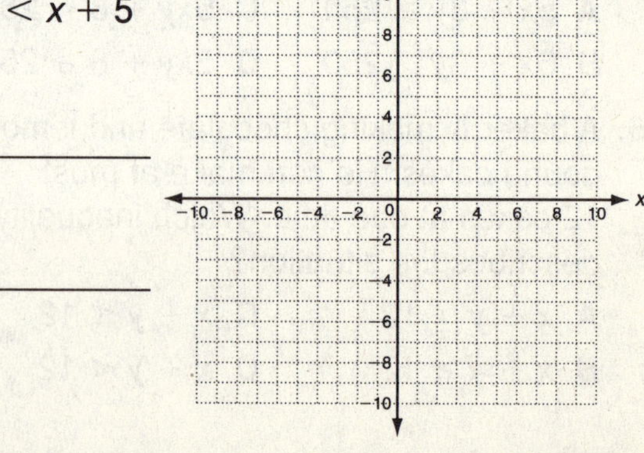

Holt Algebra 1

Practice A

LESSON 6-6

Solving Systems of Linear Inequalities

Tell whether the ordered pair is a solution of the given system.

1. $(4, 5)$; $\begin{cases} y \leq x + 2 \\ y \geq x - 1 \end{cases}$

2. $(1, 3)$; $\begin{cases} y > 3x \\ y < x + 2 \end{cases}$

3. $(2, 3)$; $\begin{cases} y < 5x - 3 \\ y \geq -x \end{cases}$

_______________   _______________   _______________

Graph the system of linear inequalities. a. Give two ordered pairs that are solutions. b. Give two ordered pairs that are not solutions.

4. $\begin{cases} y \geq x + 1 \\ y \leq -2x \end{cases}$

5. $\begin{cases} y < 2x + 4 \\ y > x - 1 \end{cases}$

6. $\begin{cases} y > -x \\ y > -x + 3 \end{cases}$

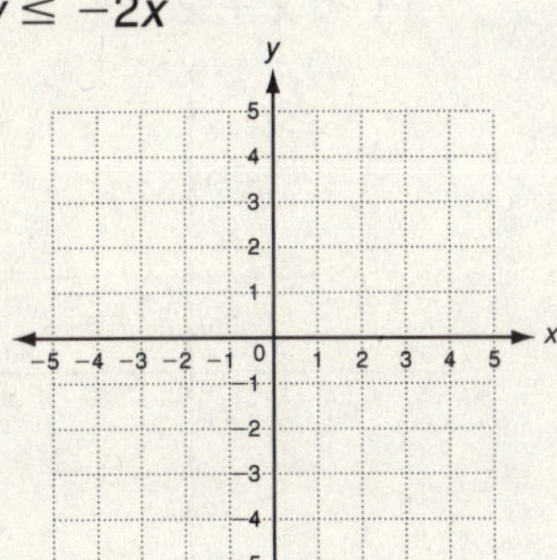
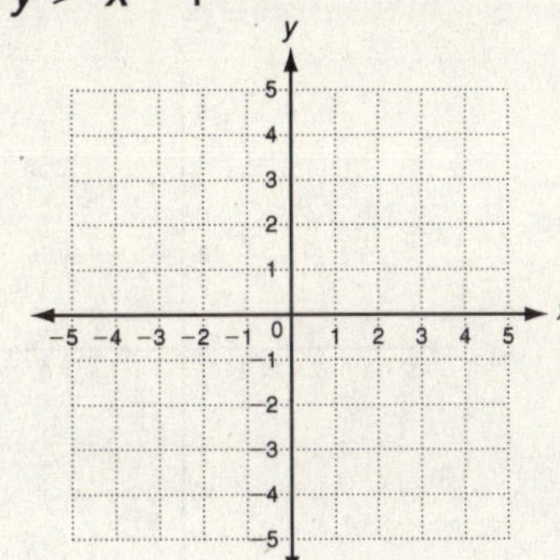
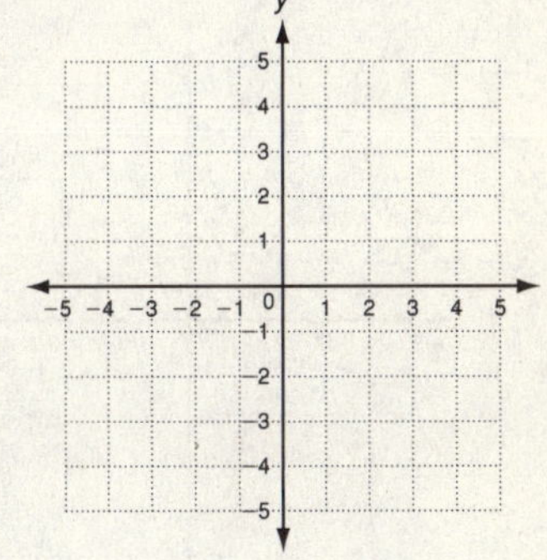

a. _______________   a. _______________   a. _______________

b. _______________   b. _______________   b. _______________

7. Lou is buying macaroni salad and potato salad for a picnic. Macaroni salad costs \$4 per pound and potato salad costs \$2 per pound. Lou would like to buy at least 6 pounds of salads and wants to spend no more than \$20.

a. Write a system of linear inequalities.
Let x = pounds of macaroni salad
Let y = pounds of potato salad

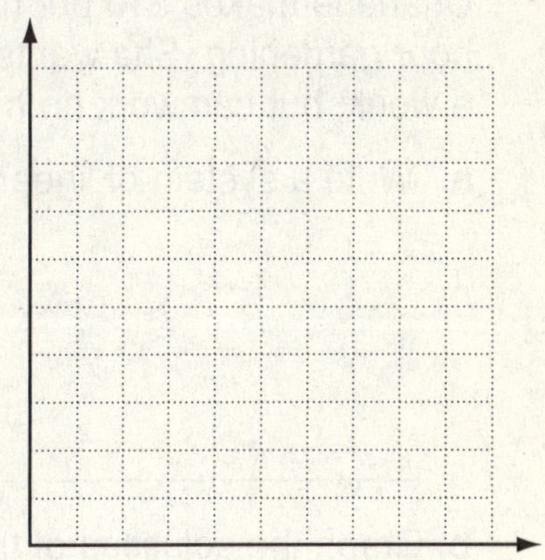

____

b. Graph the solutions of the system.

c. Describe all the possible combinations of pounds of salads that Lou could buy.

____

____

d. List two possible combinations. _______________________________

____

Holt Algebra 1

<table><tr><td>LESSON
6-6</td><td>

Practice B
Solving Systems of Linear Inequalities

</td></tr></table>

Tell whether the ordered pair is a solution of the given system.

1. $(2, -2)$; $\begin{cases} y < x - 3 \\ y > -x + 1 \end{cases}$ **2.** $(2, 5)$; $\begin{cases} y > 2x \\ y \geq x + 2 \end{cases}$ **3.** $(1, 3)$; $\begin{cases} y \leq x + 2 \\ y > 4x - 1 \end{cases}$

__________________   __________________   __________________

**Graph the system of linear inequalities. a. Give two ordered pairs
that are solutions. b. Give two ordered pairs that are not solutions.**

4. $\begin{cases} y \leq x + 4 \\ y \geq -2x \end{cases}$ **5.** $\begin{cases} y \leq \frac{1}{2}x + 1 \\ x + y < 3 \end{cases}$ **6.** $\begin{cases} y > x - 4 \\ y < x + 2 \end{cases}$

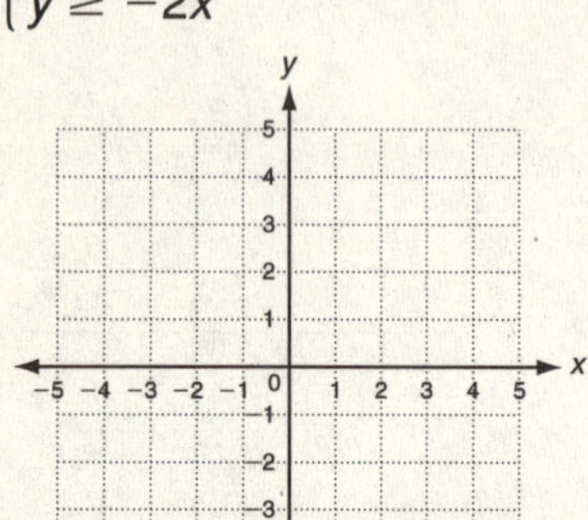   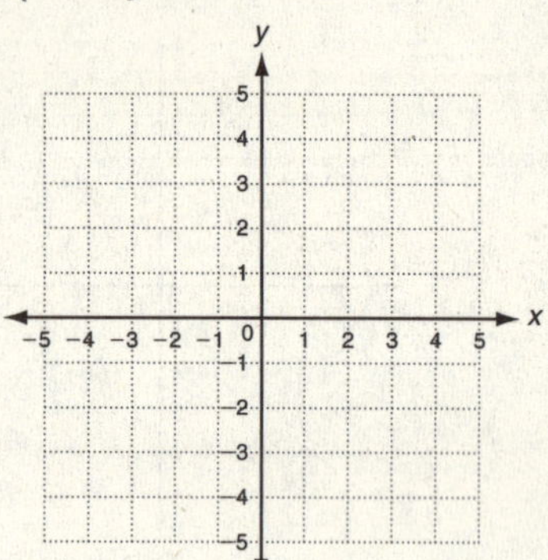   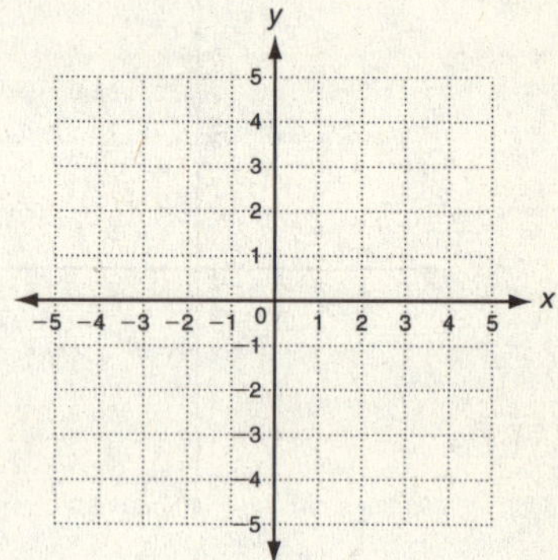

a. _______________   **a.** _______________   **a.** _______________

b. _______________   **b.** _______________   **b.** _______________

7. Charlene makes \$10 per hour babysitting and \$5 per
hour gardening. She wants to make at least \$80
a week, but can work no more than 12 hours a week.

a. Write a system of linear equations.

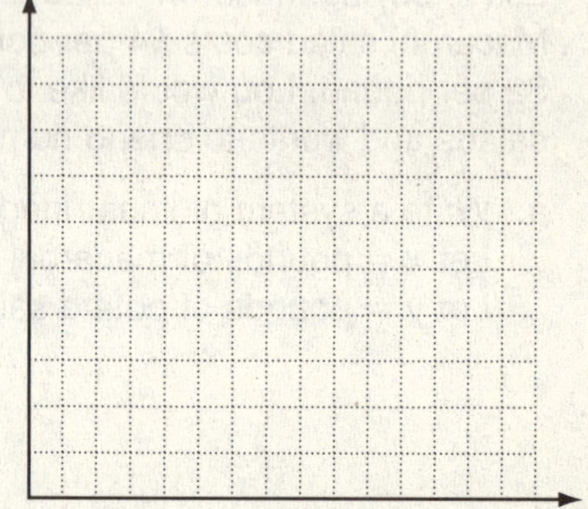

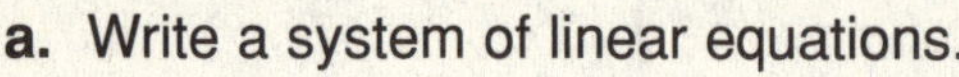

b. Graph the solutions of the system.

c. Describe all the possible combinations of hours that Charlene could work at each job.

____

____

d. List two possible combinations. _______________________________

____

     44   **Holt Algebra 1**

LESSON 6-6 — Practice C
Solving Systems of Linear Inequalities

Tell whether the ordered pair is a solution of the given system.

1. $(-2, 3);$ $\begin{cases} y \le x + 5 \\ y > -2x - 1 \end{cases}$ 2. $(-3, 3);$ $\begin{cases} y < -x + 1 \\ y > x - 4 \end{cases}$ 3. $(-1, -2);$ $\begin{cases} y > x - 2 \\ y < 3x \end{cases}$

_______________   _______________   _______________

Graph the system of linear inequalities. a. Give two ordered pairs that are solutions. b. Give two ordered pairs that are not solutions.

4. $\begin{cases} y \le 3x + 2 \\ y \ge -x \end{cases}$ 5. $\begin{cases} y > \frac{1}{3}x - 2 \\ x + y > -3 \end{cases}$ 6. $\begin{cases} 3x - 2y < 8 \\ y - 1 \le \frac{3}{2}x \end{cases}$

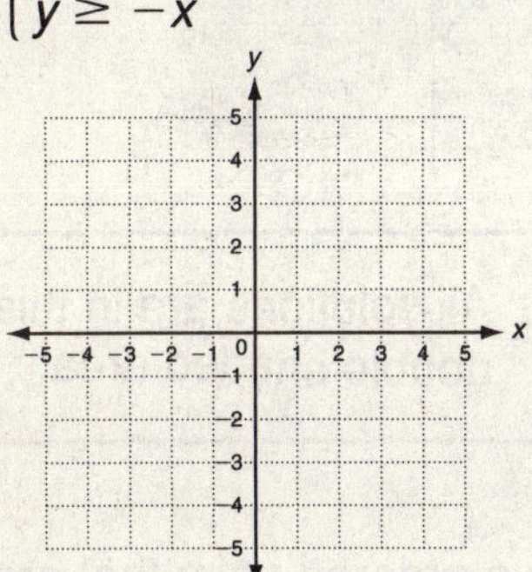   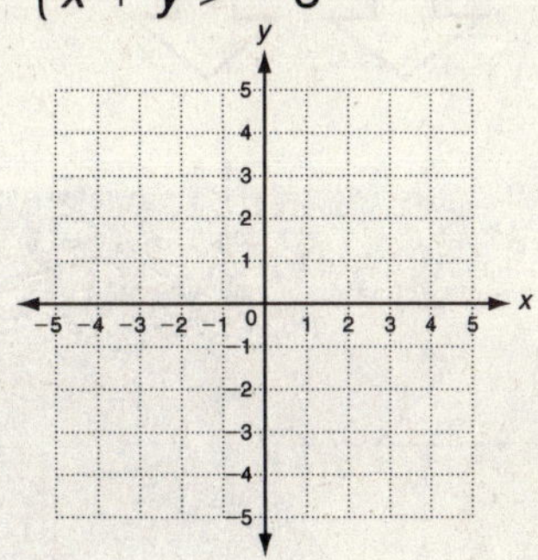   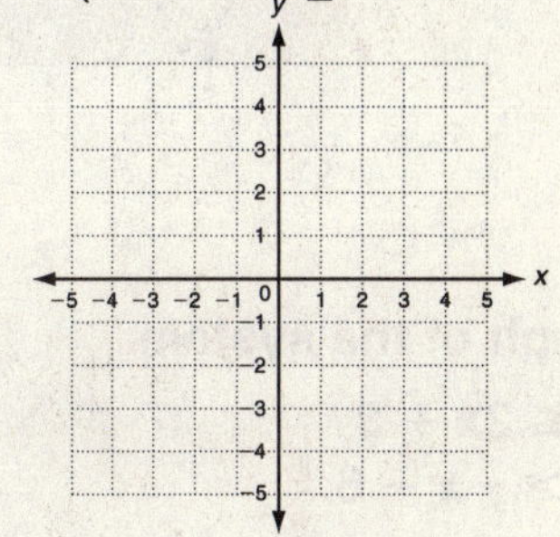

a. _______________   a. _______________   a. _______________

b. _______________   b. _______________   b. _______________

7. Dennis works at a frozen yogurt store in the summer. He needs to order boxes of small cups and boxes of large cups. The storage room can hold up to 10 more boxes of cups. Each box of small cups costs $100 and each box of large cups costs $150. A maximum of $1200 is budgeted for cups.

    a. Write a system of linear equations.

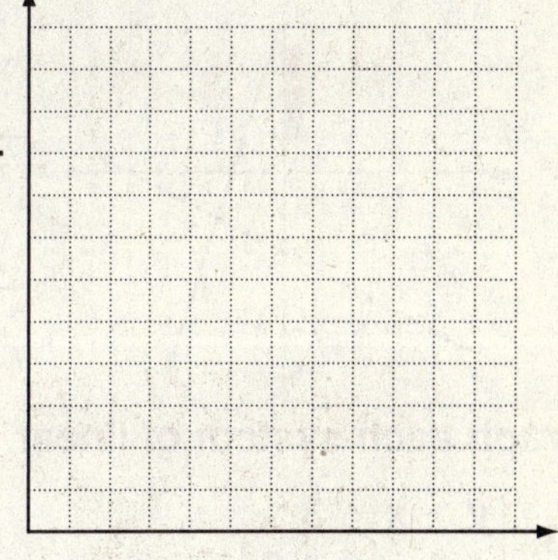

____

    b. Graph the solutions of the system.

    c. Describe all the possible combinations of boxes of cups that Dennis can order.

____

____

    d. List two possible combinations. _______________

Holt Algebra 1

Reteach
Solving Systems of Linear Inequalities

You can graph a system of linear inequalities by combining the graphs of the inequalities.

Graph of $y \leq 2x + 3$

Graph of $y > -x - 6$

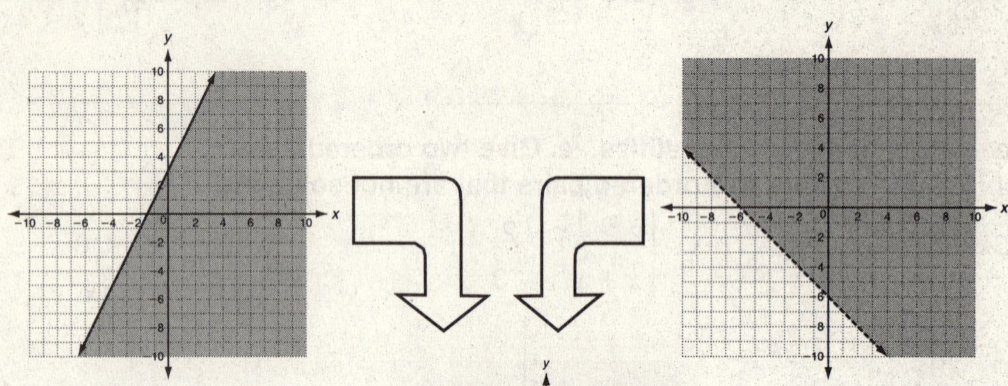

Graph of the system

$$\begin{cases} y \leq 2x + 3 \\ y > -x - 6 \end{cases}$$

All solutions are in this double shaded area.

Two ordered pairs that are solutions: $(3, 4)$ and $(5, -2)$.

For each system below, give two ordered pairs that are solutions and two that are not solutions.

1.

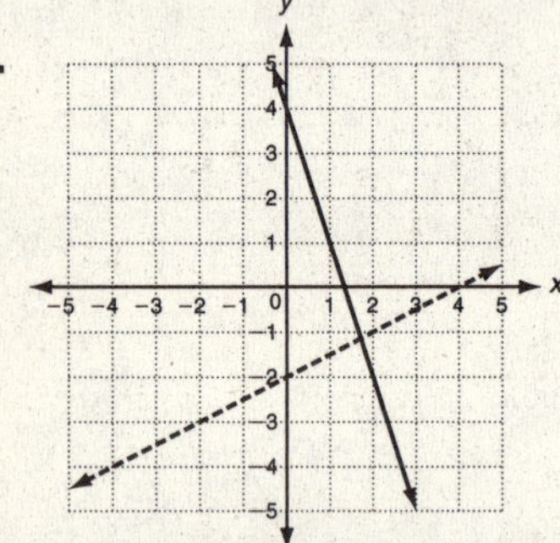

____

____

2.

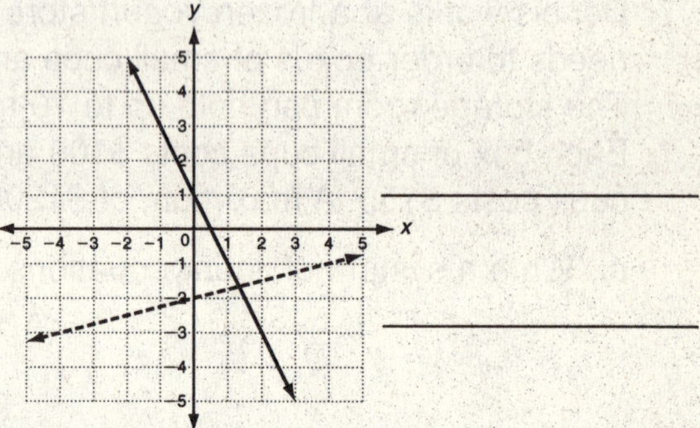

____

____

Graph each system of linear inequalities

3. $\begin{cases} y > x - 3 \\ y \geq -x + 6 \end{cases}$

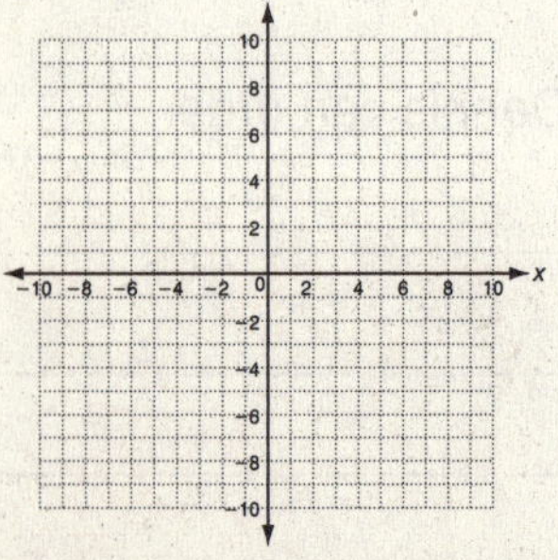

4. $\begin{cases} y < x \\ y > -2x + 1 \end{cases}$

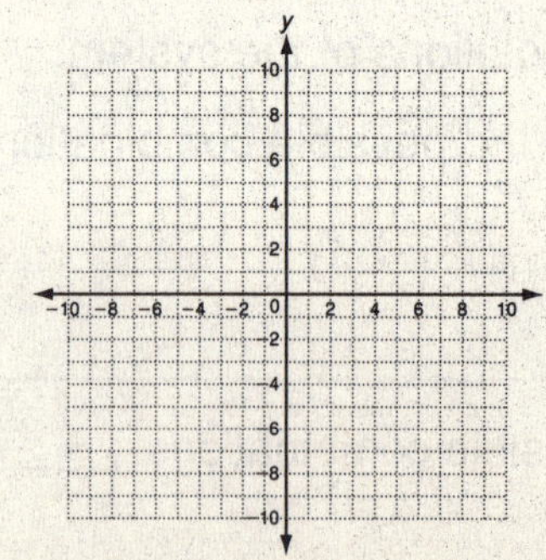

Holt Algebra 1

Reteach

LESSON 6-6

Solving Systems of Linear Inequalities (continued)

A system of equations with parallel lines has no solutions.
Parallel lines in a system of inequalities might have solutions.

Graph $\begin{cases} y < x + 4 \\ y > x - 2 \end{cases}$.

Graph $\begin{cases} y \geq 2x + 4 \\ y > 2x - 1 \end{cases}$.

Graph $\begin{cases} y > -3x + 5 \\ y < -3x - 3 \end{cases}$.

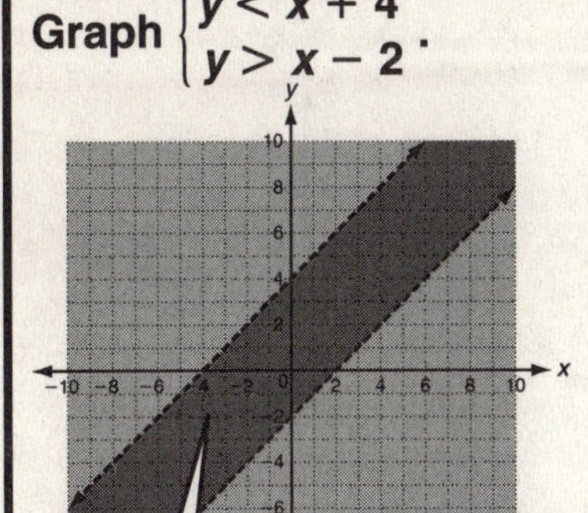

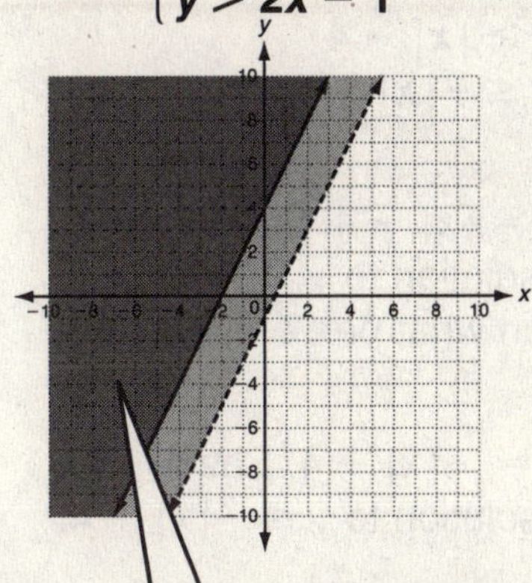

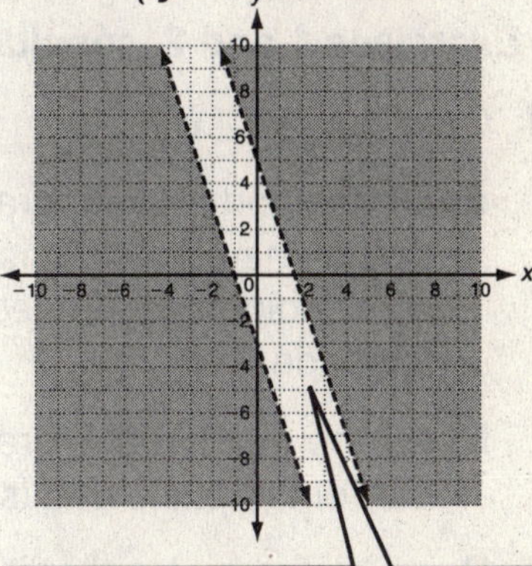

Solutions are in the double shaded area between the parallel lines.

Solutions are in the double shaded area to one side of the line.

There are no overlapping areas. There are no solutions.

Graph the solutions of each linear inequality.

5. $\begin{cases} y \leq x - 3 \\ y > x + 3 \end{cases}$

6. $\begin{cases} y > 2x - 2 \\ y \leq 2x + 3 \end{cases}$

7. $\begin{cases} y > -x - 1 \\ y > -x - 5 \end{cases}$

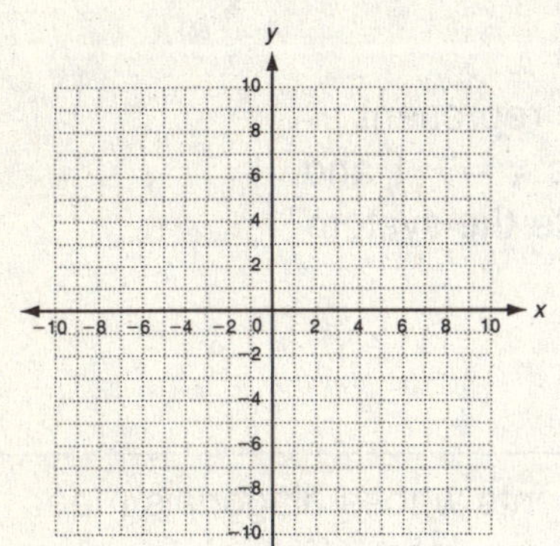

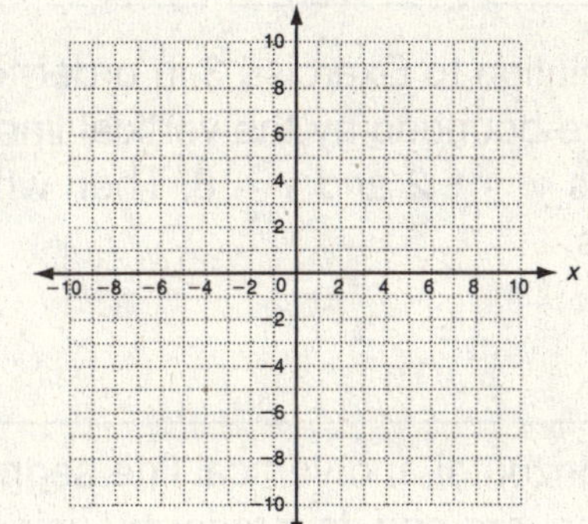

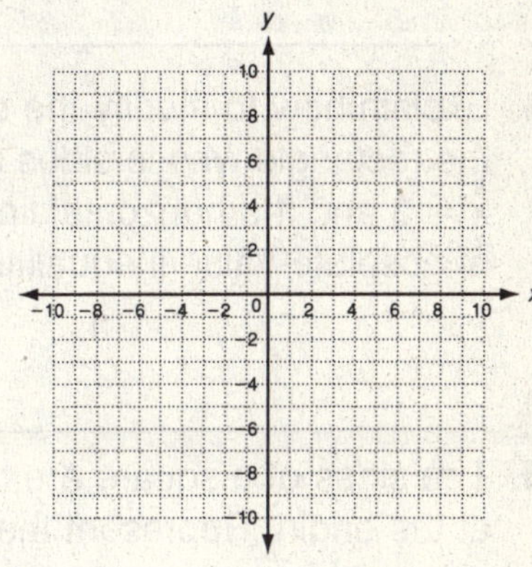

Holt Algebra 1

Challenge
Solving Systems of Linear Inequalities

Recall that $|x|$ has two parts in its definition. If $x \geq 0$, then $|x| = x$. If $x < 0$, then $|x| = -x$. This definition is useful when dealing with systems of inequalities that involve absolute value.

In Exercises 1 and 2, consider $\begin{cases} y \geq |x| - 4 \\ y \leq -|x| + 4 \end{cases}$.

1. Consider $y \geq |x| - 4$.

   a. On the grid at right, graph $y = |x| - 4$. Shade the part of the plane that is the solution to $y \geq |x| - 4$. (Test points to help decide on what part of the plane to shade.)

   b. On the grid at right, graph $y = -|x| + 4$. Shade that part of the plane that is the solution to $y \leq -|x| + 4$.

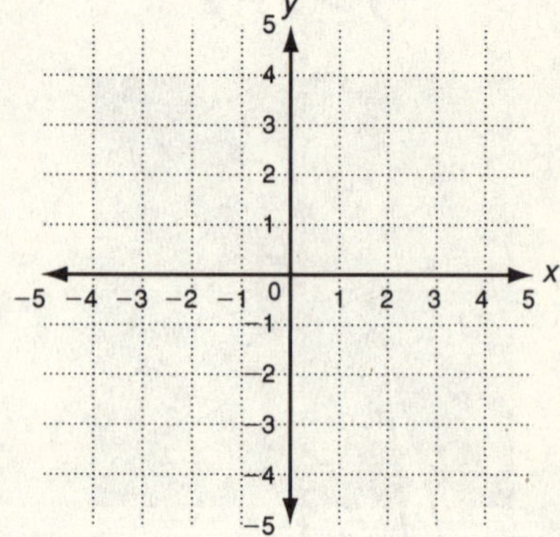

2. Describe the region that is the common solution to the two absolute-value inequalities.

____

You can reverse the process and represent a specified region by using a system of inequalities.

3. On the grid at right, graph the vertical lines $x = 3$ and $x = -3$. Also graph the horizontal lines $y = 5$ and $y = -5$.

   a. Write a pair of absolute-value inequalities whose graphs are the interior of the rectangle formed.

   ____

   b. Write a pair of absolute-value inequalities whose graphs are the exterior of the rectangle formed.

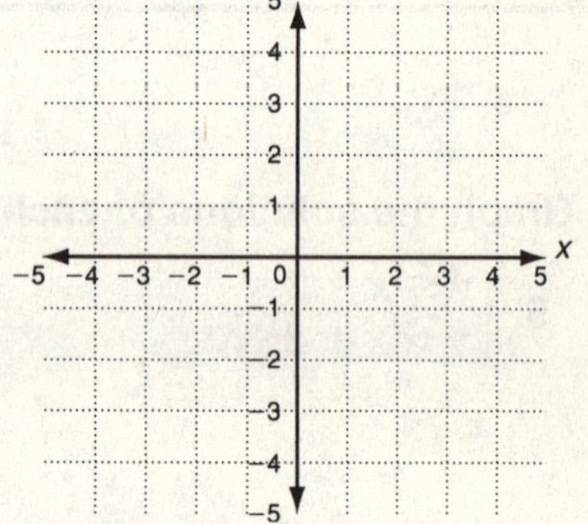

   ____

4. Explain how to modify the solution to Exercise 3 in order to represent the rectangle whose sides are bounded by the vertical lines $x = -1$ and $x = 3$ and the horizontal lines $y = -2$ and $y = 8$. Then write the system of absolute-value inequalities.

____

5. The sides of a square are horizontal and vertical line segments whose diagonals meet at the origin. Represent the square and its interior by using a system of absolute-value inequalities.

____

Holt Algebra 1

Problem Solving
Solving Systems of Linear Inequalities

Write the correct answer.

1. Paul earns $7 per hour at the bagel shop and $12 per hour mowing lawns. Paul needs to earn at least $120 per week, but he must work less than 30 hours per week. Write and graph the system of linear inequalities that describes this situation.

____

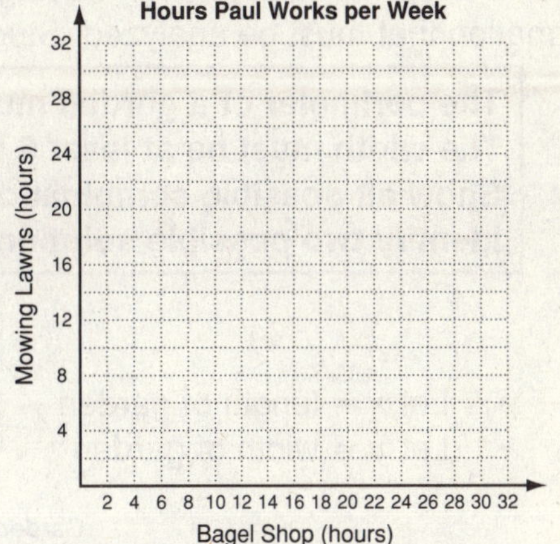

2. Zoe plans to knit a scarf. She wants the scarf to be more than 1 but less than 1.5 feet wide, and more than 6 but less than 8 feet long. Graph all possible dimensions of Zoe's scarf. List two possible combinations.

____

____

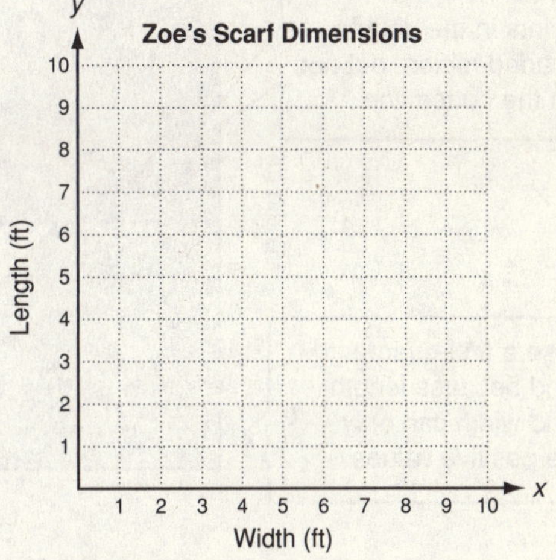

The graph shows the numbers of two types of custom wood tables that can be made to fit a client's needs. Select the best answer.

3. Which system of linear inequalities represents the graph?

A $\begin{cases} x + y \le 15 \\ y \ge 12 - \frac{4}{3}x \end{cases}$ C $\begin{cases} x + y \ge 15 \\ y \ge \frac{4}{3}x - 12 \end{cases}$

B $\begin{cases} y \le x + 15 \\ y \ge 12 - \frac{4}{3}x \end{cases}$ D $\begin{cases} y \le 15 - x \\ y \le \frac{4}{3}x - 12 \end{cases}$

4. If 6 buffet tables are built, which can NOT be the number of dining tables built?

F 4   H 8

G 6   J 10

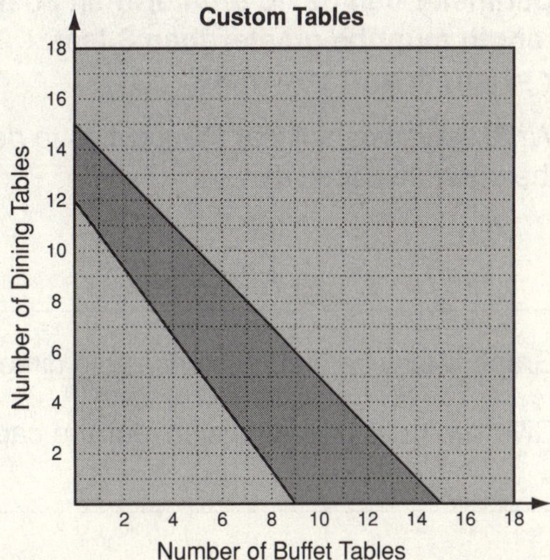

Holt Algebra 1

Reading Strategies

LESSON 6-6

Analyze Information

To solve a real-world problem involving inequalities, there is a lot of information that must be analyzed. Study the example below.

> **The perimeter of a garden must be less than 20 meters.**
> **The width must be at least 6 meters.**
> **Show all possible combinations of the garden's length and width.**
> **Identify two possible solutions.**

Define variables for the unknowns: "the garden's length and width."

The perimeter must be less than 20 meters: $P = 2l + 2w$; and "less than" means $<$.

- Let x = length of garden
- Let y = width of garden

$$\begin{cases} 2x + 2y < 20 \\ y \geq 6 \end{cases}$$

The width must be at least 6 meters: y = width and "at least" means $\geq$.

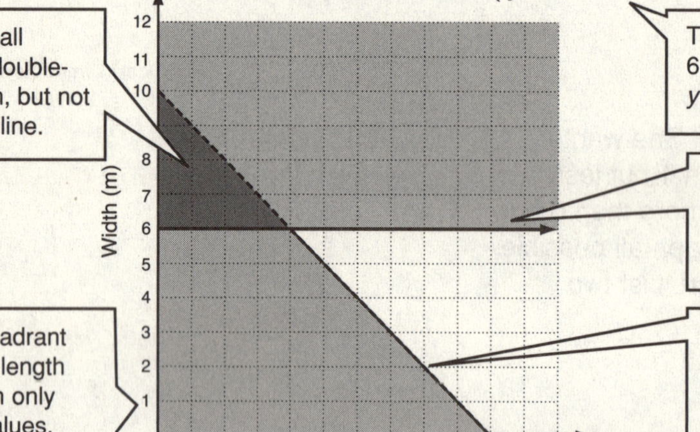

Solutions are all points in the double-shaded region, but not on the dotted line.

$y \geq 6$
solid line
shaded above

Use a first-quadrant grid because length and width can only be positive values.

Rewrite $2x + 2y < 20$ in slope-intercept form:
$$y < -x + 10$$
dotted line, shaded below

Possible solutions: l = 2 m, w = 7 m and l = 1 m, w = 8 m

Manuel is going to build a rabbit cage with a rectangular base.
The perimeter can be no greater than 30 feet.
The length must be greater than 8 feet.
Let x = length and y = width.

1. Write a system of linear inequalities to describe the possible cage sizes.

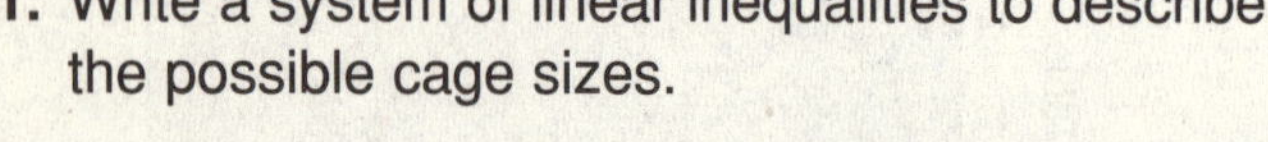

2. Graph the system to show possible dimensions.

3. Give two possible dimensions for the cage.

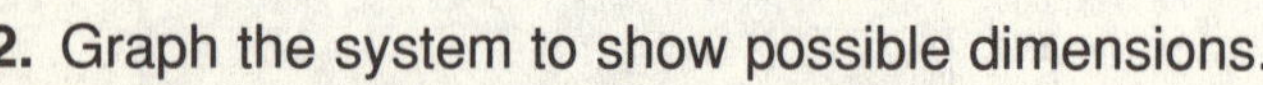

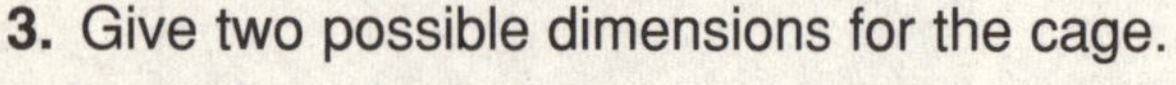

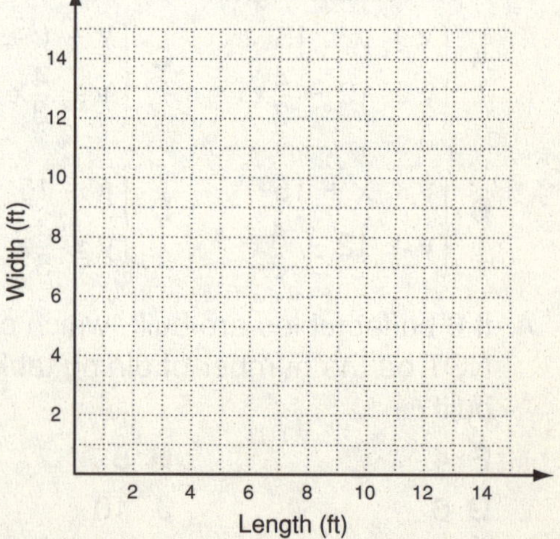

Holt Algebra 1

Practice A
Solving Systems by Graphing

Complete the steps to determine whether the ordered pair is a solution of the given system. Circle ✓ or ✗ for each equation. Then, write _is_ or _is not_ to complete the sentence.

1. $(2, 4)$; $\begin{cases} x - y = -2 \\ 2x + y = 6 \end{cases}$

2. $(1, -2)$; $\begin{cases} 2x + y = 0 \\ x + 4y = -7 \end{cases}$

$x - y = -2$		$2x + y = 6$	
$(2) - (4)$	-2	$2(2) + \boxed{4}$	6
-2	-2	$\boxed{4} + \boxed{4}$	6
		$\boxed{8}$	6
✓ or ✗		✓ or ✗	

$2x + y = 0$		$x + 4y = -7$	
$2(1) + (-2)$	0	$(1) + 4\boxed{-2}$	-7
$\boxed{2} + (-2)$	0	$\boxed{1} + \boxed{-8}$	-7
$\boxed{0}$	0	$\boxed{-7}$	-7
✓ or ✗		✓ or ✗	

$(2, 4)$ __is not__ a solution of the system.

$(1, -2)$ __is__ a solution of the system.

Solve each system by graphing. One of the lines has been graphed for you.

3. $\begin{cases} y = 3x - 5 \\ y = x - 3 \end{cases}$ Solution: __$(1, -2)$__

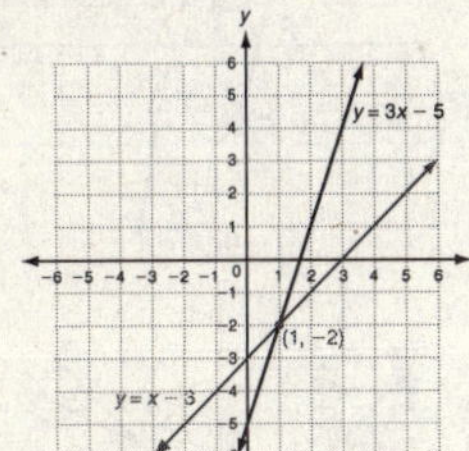

4. $\begin{cases} y = x + 7 \\ y = -2x - 2 \end{cases}$ Solution: __$(-3, 4)$__

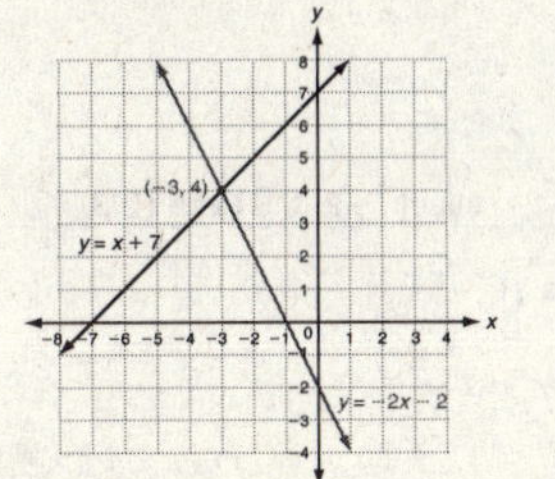

5. The Science Club needs to rent a bus for a field trip. Main Street Buses charges a $40 rental fee, plus $2 per mile. County Bus Line charges a $20 rental fee, plus $3 per mile. For what number of miles will the total charge be the same? What will that charge be?

__20 miles; $80__

Practice B
Solving Systems by Graphing

Tell whether the ordered pair is a solution of the given system.

1. $(3, 1)$; $\begin{cases} x + 3y = 6 \\ 4x - 5y = 7 \end{cases}$ __yes__

2. $(6, -2)$; $\begin{cases} 3x - 2y = 14 \\ 5x - y = 32 \end{cases}$ __no__

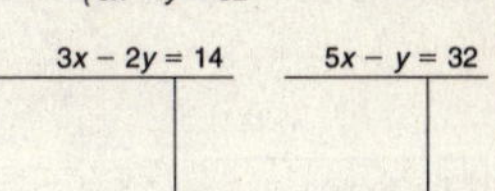

$x + 3y = 6$	$4x - 5y = 7$	$3x - 2y = 14$	$5x - y = 32$

Solve each system by graphing. Check your answer.

3. $\begin{cases} y = x + 4 \\ y = -2x + 1 \end{cases}$ Solution: __$(-1, 3)$__

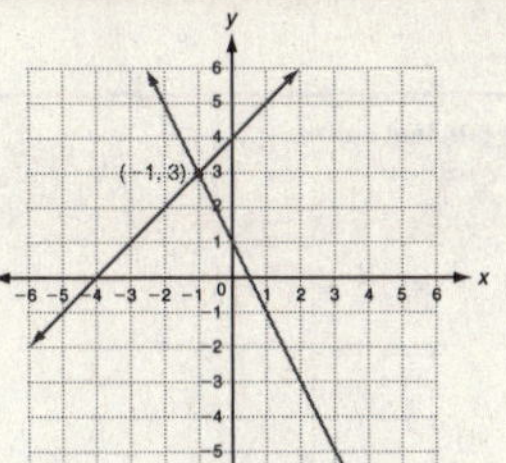

4. $\begin{cases} y = x + 6 \\ y = -3x + 6 \end{cases}$ Solution: __$(0, 6)$__

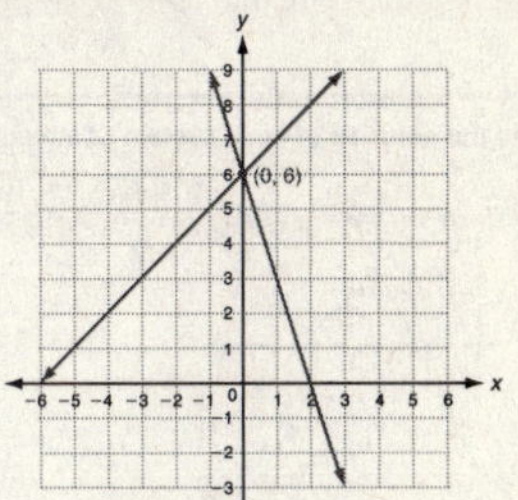

5. Maryann and Carlos are each saving for new scooters. So far, Maryann has $9 saved, and can earn $6 per hour babysitting. Carlos has $3 saved, and can earn $9 per hour working at his family's restaurant. After how many hours of work will Maryann and Carlos have saved the same amount? What will that amount be?

__2 hours; $21__

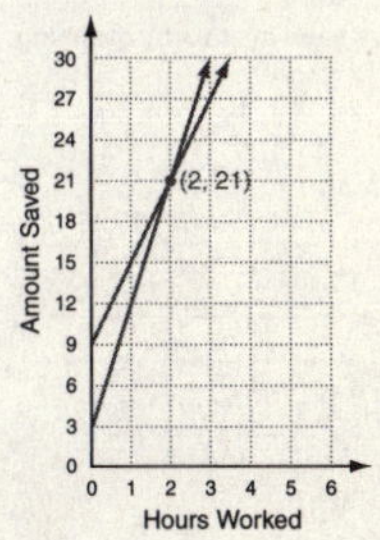

Practice C
Solving Systems by Graphing

Tell whether the ordered pair is a solution of the given equation.

1. $(6, -2)$; $\begin{cases} 2x - y = 14 \\ x + 4y = -2 \end{cases}$ __yes__

2. $(4, 0)$; $\begin{cases} x - 2y = 4 \\ -x + y = -8 \end{cases}$ __no__

3. $(-6, -2)$; $\begin{cases} 2x - y = -10 \\ -x + y = 4 \end{cases}$ __yes__

Solve each system by graphing.

4. $\begin{cases} y = 2x + 4 \\ y = -x + 7 \end{cases}$

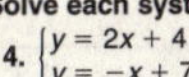
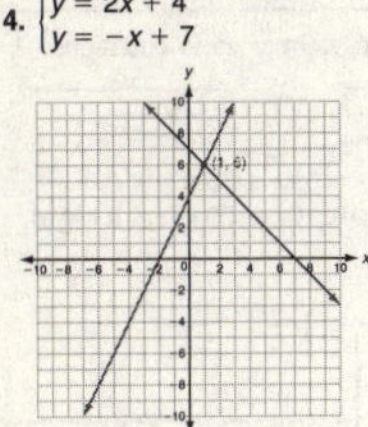

Solution: __$(1, 6)$__

5. $\begin{cases} y = 2x - 6 \\ y = 3x - 8 \end{cases}$

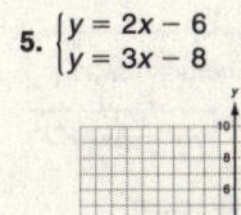
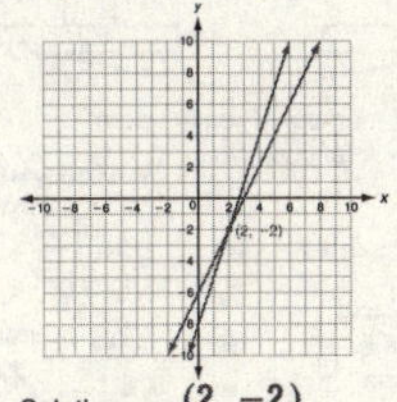

Solution: __$(2, -2)$__

6. $\begin{cases} x + y = -2 \\ y = 4x - 7 \end{cases}$

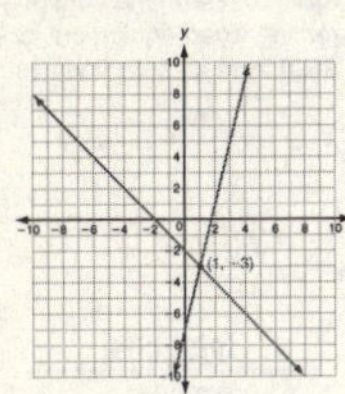

Solution: __$(1, -3)$__

7. $\begin{cases} x = y + 2 \\ 2x = y \end{cases}$

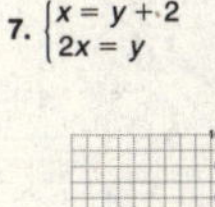
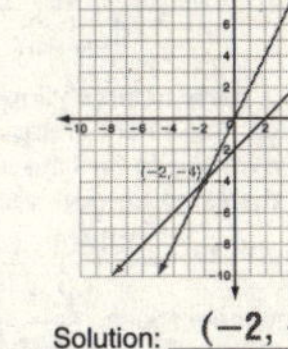

Solution: __$(-2, -4)$__

Use a graphing calculator to solve.

8. To sell an item in an online auction, WebAuctions charges a $5 listing fee plus 10% of the final selling price. AuctionsOnline charges a $3 listing fee plus 15% of the final selling price. For what final selling price do both companies charge the same amount? What will that amount be? __$40; $9__

Reteach
Solving Systems by Graphing

You have checked to see if an ordered pair was a solution of an equation. Now you will check to see if an ordered pair is a solution of a system of equations.

Tell whether $(1, 9)$ is a solution of
$$\begin{cases} x + y = 10 \\ 3x + y = 12 \end{cases}$$

Step 1: Substitute $(1, 9)$ into one of the equations.

$(1, 9)$ means that $x = 1$ and $y = 9$.

$x + y \overset{?}{=} 10$

$1 + 9 \overset{?}{=} 10$

$10 \overset{?}{=} 10$ ✓

Solution checks. Continue with Step 2

Step 2: Substitute $(1, 9)$ into the other equation.

$3x + y = 12$

$3(1) + 9 \overset{?}{=} 12$

$3 + 9 \overset{?}{=} 12$

$12 \overset{?}{=} 12$ ✓

The ordered pair makes both equations true. So $(1, 9)$ is a solution of the system.

Tell whether $(2, -3)$ is a solution of
$$\begin{cases} x + y = 5 \\ 2x + 5y = -11 \end{cases}$$

Step 1: Substitute $(2, -3)$ into one of the equations.

$x + y = 5$

$2 + -3 \overset{?}{=} 5$

$-1 \overset{?}{=} 5$ ✗

Stop! There is no need to check the other equation. The ordered pair is not a solution of the system.

Tell whether the ordered pair is a solution of the given system.

1. $(0, -4)$; $\begin{cases} x + 2y = -8 \\ x = 4 + y \end{cases}$ __yes__

2. $(2, 5)$; $\begin{cases} x + y = 7 \\ 3x + y = 10 \end{cases}$ __no__

3. $(-3, 1)$; $\begin{cases} 2x + y = 5 \\ x + 3y = -6 \end{cases}$ __no__

4. $(-3, 9)$; $\begin{cases} y = x + 12 \\ y = -3x \end{cases}$ __yes__

Reteach
Solving Systems by Graphing (continued)

Graph to check if $(5, 7)$ is a solution of $\begin{cases} y = x + 2 \\ y = 2x + 3 \end{cases}$.

If $(5, 7)$ is not the solution, find the solution from the graph.

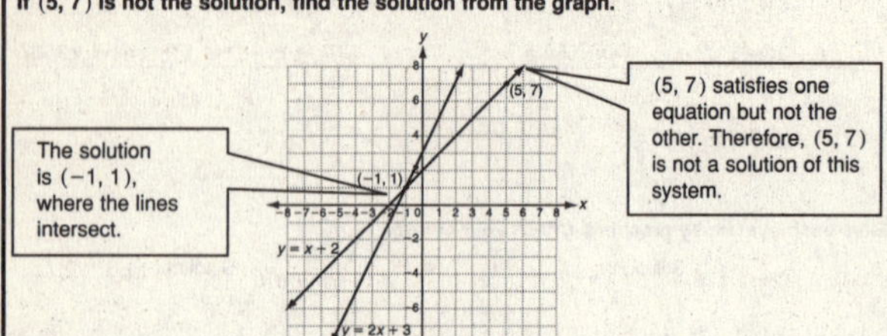

Find the solution of each system of equations graphed below.

5.
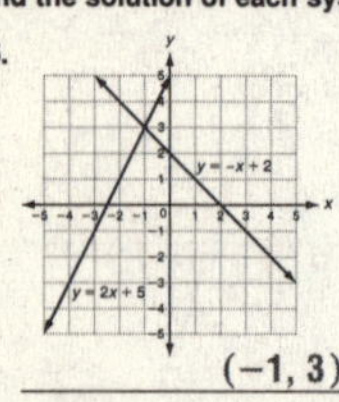

$(-1, 3)$

6.
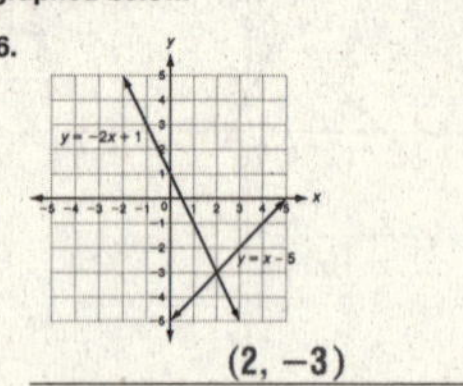

$(2, -3)$

Solve each system by graphing.

7. $\begin{cases} y = -3 \\ y = x + 2 \end{cases}$

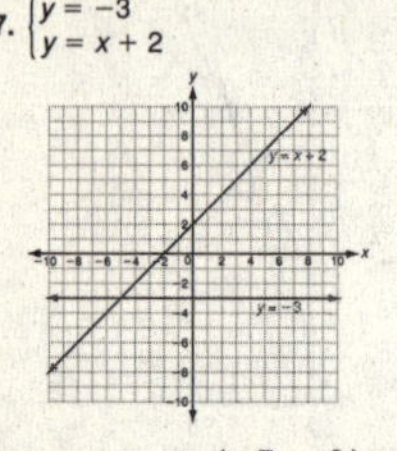

$(-5, -3)$

8. $\begin{cases} y = x - 6 \\ y = -x \end{cases}$

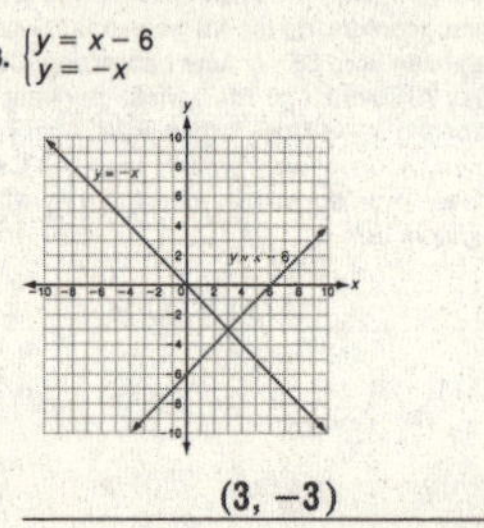

$(3, -3)$

7

Holt Algebra 1

Challenge
Solving Systems by Graphing

Sometimes one or both equations in a system will be non-linear. The solutions to these systems will still be where the graphs intersect. This can happen in more than one place. Recall that you can graph any equation by generating and plotting ordered pairs.

Solve each system by graphing.

1. $\begin{cases} y = x^2 - 4 \\ y = x + 2 \end{cases}$

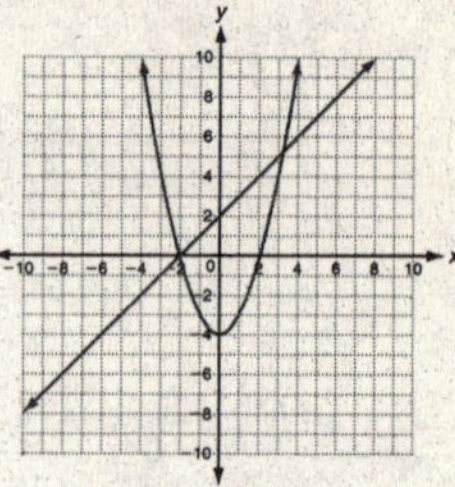

$(-2, 0)$ and $(3, 5)$

2. $\begin{cases} y = -x^2 + 5 \\ y = x^2 - 5 \end{cases}$

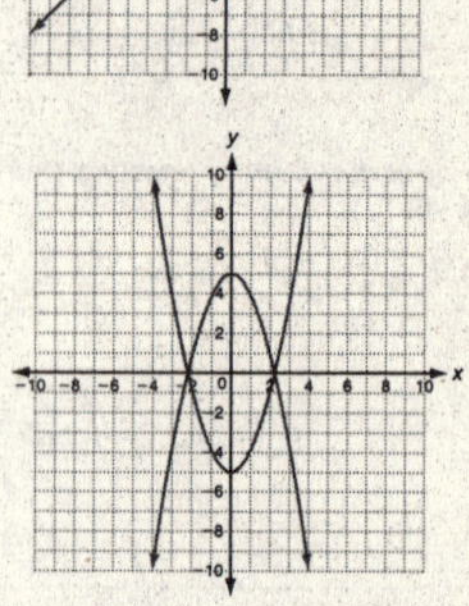

about $(-2.3, 0)$ and $(2.3, 0)$

3. $\begin{cases} y = |x + 1| \\ y = \frac{1}{2}x + 3 \end{cases}$

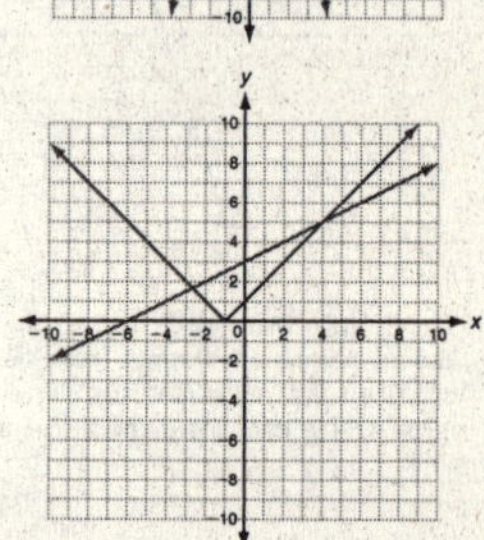

about $(-2.8, 1.9)$ and $(3.9, 4.9)$

8

Holt Algebra 1

Problem Solving
Solving Systems by Graphing

Write the correct answer.

1. Mr. Malone is putting money in two savings accounts. Account A started with $200 and Account B started with $300. Mr. Malone deposits $15 in Account A and $10 in Account B each month. In how many months will the accounts have the same balance? What will that balance be?

20 months, $500

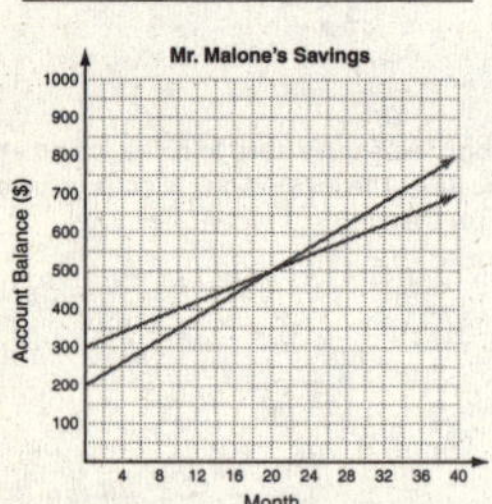

2. Tom currently has 5 comic books in his collection and has subscribed to receive 5 new comic books each month. His uncle has 145 comic books, but sends 5 to each of his 3 nieces each month. In how many months will they have the same number of comic books? How many books will that be?

7 months, 40 books

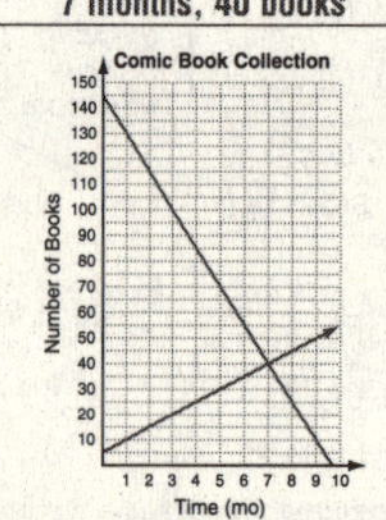

The graph below compares the heights of two trees. Use the graph to answer questions 3–5. Select the best answer.

3. How many years after planting will the trees be the same height?
A 1 years   **C 4 years**
B 2 years   D 6 years

4. Which system of equations is represented by the graph?
F $\begin{cases} y = x + 2 \\ y = 0.5x + 2 \end{cases}$ H $\begin{cases} y = 2x + 4 \\ y = x + 4 \end{cases}$
G $\begin{cases} y = x + 2 \\ y = 0.5x + 4 \end{cases}$ J $\begin{cases} y = 4x - 2 \\ y = 2x + 2 \end{cases}$

5. How fast does the tree that started at 2 feet tall grow?
A 0.5 ft/yr   C 1.5 ft/yr
B 1 ft/yr   D 2 ft/yr

6. How fast does the tree that started at 4 feet tall grow?
F 0.5 ft/yr   H 1.5 ft/yr
G 1 ft/yr   J 2 ft/yr

9

Holt Algebra 1

Reading Strategies
Follow a Procedure

In the example below, the steps show how to set up and solve a system of linear equations.

A teacher needs to rent video equipment. Rent All charges $60 plus $12 for each hour. Rent It Here charges $80 plus $10 per hour. After how many hours is the cost of renting video equipment the same from both companies?

1. Define variables for the unknowns.

2. Write a system of linear equations.

- Let x = number of rental hours
- Let y = total cost

$\begin{cases} y = 12x + 60 \text{ (Rent All)} \\ y = 10x + 80 \text{ (Rent It Here)} \end{cases}$

3. Graph the equations on the same grid.

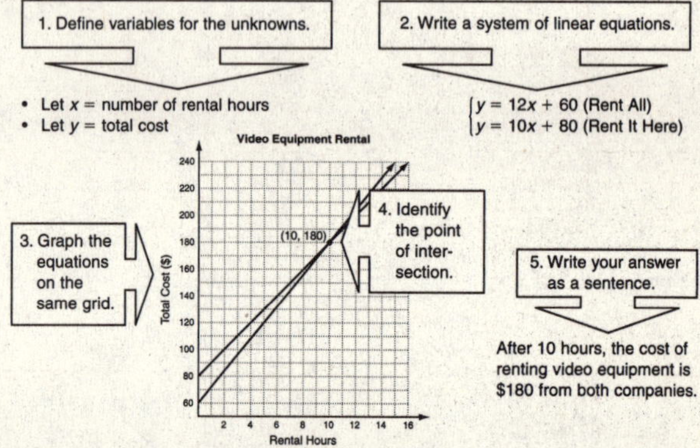

4. Identify the point of intersection.

5. Write your answer as a sentence.

After 10 hours, the cost of renting video equipment is $180 from both companies.

Derek is going to have a party catered. Good Eats charges $120 plus $10 per person. Food Fare charges $150 plus $8 per person. Complete the following to find the number of people for which the total cost is the same for both catering companies.

1. Define the variables. Let x = **number of people**, Let y = **total cost**

2. Write a system of linear equations. $\begin{cases} y = 10x + 120 \\ y = 8x + 150 \end{cases}$

3. Graph the equations.

4. What is the point of intersection? **(15, 270)**

5. Write your answer. **For 15 people, the total cost will be $270 with both catering companies.**

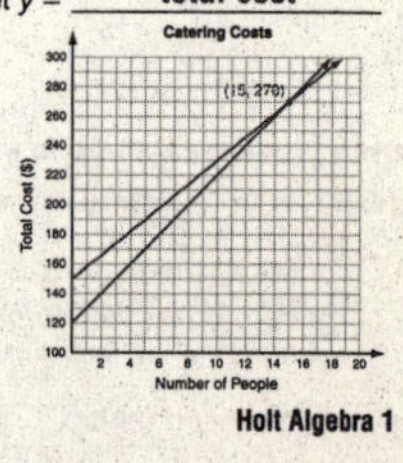

10

Holt Algebra 1

Holt Algebra 1

Practice A
Solving Systems by Substitution

Fill in the blanks to solve each system by substitution.

1. $\begin{cases} y = 3x \\ y = x + 4 \end{cases}$

Substitute ____$3x$____ for y in the second equation.

$\underline{3x} = x + 4$

$-\underline{x} \quad - x$

$\underline{2x} = 4$

$\div \underline{2} \quad \div \underline{2}$

$\underline{x} = \underline{2}$

Since $x = \underline{2}$, substitute $\underline{2}$ for x in one of the equations to find the value of y:

$y = 3x$

$y = 3(\underline{2})$

$y = \underline{6}$

Solution: $(\underline{2}, \underline{6})$

2. $\begin{cases} 3x + y = 25 \\ y = x - 3 \end{cases}$

Substitute ____$x - 3$____ for y in the first equation.

$3x + (\underline{x - 3}) = 25$

$\underline{4x} - 3 = 25$

$ +3 \quad +3$

$\underline{4x} = 28$

$\div \underline{4} \quad \div \underline{4}$

$\underline{x} = \underline{7}$

Since $x = \underline{7}$, substitute $\underline{7}$ for x in one of the equations to find the value of y:

$y = x - 3$

$y = \underline{7} - 3$

$y = \underline{4}$

Solution: $(\underline{7}, \underline{4})$

Solve each system by substitution. Check your answer.

3. $\begin{cases} y = 4x \\ y = 2x + 6 \end{cases}$ $\underline{(3, 12)}$

4. $\begin{cases} y = x - 2 \\ 2x + y = 4 \end{cases}$ $\underline{(2, 0)}$

5. $\begin{cases} 2x + y = -1 \\ -x + y = -7 \end{cases}$ $\underline{(2, -5)}$

6. A professional organizer charges a $45 consultation fee, plus $20 per hour. Her competitor charges a $30 consultation fee, plus $26 per hour.

  a. Write a system of equations to represent the situation. $\underline{\begin{cases} y = 20x + 45 \\ y = 26x + 30 \end{cases}}$

  b. Solve the system by substitution. For what number of hours will the total charge be the same? ____2.5____

  c. What will that charge be? ____95____

Holt Algebra 1

Practice B
Solving Systems by Substitution

Solve each system by substitution. Check your answer.

1. $\begin{cases} y = x - 2 \\ y = 4x + 1 \end{cases}$ $\underline{(-1, -3)}$

2. $\begin{cases} y = x - 4 \\ y = -x + 2 \end{cases}$ $\underline{(3, -1)}$

3. $\begin{cases} y = 3x + 1 \\ y = 5x - 3 \end{cases}$ $\underline{(2, 7)}$

4. $\begin{cases} 2x - y = 6 \\ x + y = -3 \end{cases}$ $\underline{(1, -4)}$

5. $\begin{cases} 2x + y = 8 \\ y = x - 7 \end{cases}$ $\underline{(5, -2)}$

6. $\begin{cases} 2x + 3y = 0 \\ x + 2y = -1 \end{cases}$ $\underline{(3, -2)}$

7. $\begin{cases} 3x - 2y = 7 \\ x + 3y = -5 \end{cases}$ $\underline{(1, -2)}$

8. $\begin{cases} -2x + y = 0 \\ 5x + 3y = -11 \end{cases}$ $\underline{(-1, -2)}$

9. $\begin{cases} \frac{1}{2}x + \frac{1}{3}y = 5 \\ \frac{1}{4}x + y = 10 \end{cases}$ $\underline{(4, 9)}$

Write a system of equations to represent the situation. Then, solve the system by substitution.

10. The length of a rectangle is 3 more than its width. The perimeter of the rectangle is 58 cm. What are the rectangle's dimensions?

$\underline{\begin{cases} l = w + 3 \\ 2l + 2w = 58 \end{cases}}$; 13 cm by 16 cm

11. Carla and Benicio work in a men's clothing store. They earn commission from each suit and each pair of shoes they sell. For selling 3 suits and one pair of shoes, Carla has earned $47 in commission. For selling 7 suits and 2 pairs of shoes, Benicio has earned $107 in commission. How much do the salespeople earn for the sale of a suit? for the sale of a pair of shoes?

$\underline{\begin{cases} 3s + 1p = 47 \\ 7s + 2p = 107 \end{cases}}$; suit: $13; pair of shoes: $8

Holt Algebra 1

Practice C
Solving Systems by Substitution

Solve each system by substitution. Check your answer.

1. $\begin{cases} y = x - 6 \\ y = 2x - 8 \end{cases}$ $\underline{(2, -4)}$

2. $\begin{cases} y = x - 4 \\ y = -2x + 5 \end{cases}$ $\underline{(3, -1)}$

3. $\begin{cases} y = 2x + 8 \\ y = 8 \end{cases}$ $\underline{(0, 8)}$

4. $\begin{cases} x + y = -3 \\ x - y = -1 \end{cases}$ $\underline{(-2, -1)}$

5. $\begin{cases} 4x + 2y = -2 \\ y = 6x - 5 \end{cases}$ $\underline{\left(\frac{1}{2}, -2\right)}$

6. $\begin{cases} \frac{1}{2}x = y - 1 \\ \frac{1}{3}x = y \end{cases}$ $\underline{(-6, -2)}$

7. $\begin{cases} 4x + y = 3.4 \\ x = 4y \end{cases}$ $\underline{(0.8, 0.2)}$

8. $\begin{cases} x = -\frac{1}{4}y + 5 \\ 3x + 2y = 0 \end{cases}$ $\underline{(8, -12)}$

9. $\begin{cases} 3x = -y + 10 \\ 2x + 3y = -12 \end{cases}$ $\underline{(6, -8)}$

Solve.

10. The sum of two numbers is 39. The greater number is 5 less than 3 times the smaller number. What are the two numbers?

    11 and 28

11. Adam has dimes and quarters in his pocket. There are 28 coins in all, and they are worth $4.60 altogether. How many quarters does Adam have?

    12 quarters

12. At a pet store, Rhonda paid $11.50 for 3 dog bowls and 4 bones. Kelly paid $13 for 2 dog bowls and 8 bones. How much will Andrew pay for 4 dog bowls and 2 bones?

    $12.00

Holt Algebra 1

Reteach
Solving Systems by Substitution

You can use substitution to solve a system of equations if one of the equations is already solved for a variable.

Solve $\begin{cases} y = x + 2 \\ 3x + y = 10 \end{cases}$

Step 1: Choose the equation to use as the substitute.

Use the first equation $y = x + 2$ because it is already solved for a variable.

Step 2: Solve by substitution.

$3x + y = 10$

$3x + (x + 2) = 10$ *Substitute $x + 2$ for y.*

$4x + 2 = 10$ *Combine like terms.*

$\quad -2 \quad -2$

$4x = 8$

$\dfrac{4x}{4} = \dfrac{8}{4}$

$x = 2$

Step 3: Now substitute $x = 2$ back into one of the original equations to find the value of y.

$y = x + 2$

$y = 2 + 2$

$y = 4$

The solution is $(2, 4)$.

Check:

Substitute $(2, 4)$ into both equations.

$y = x + 2 \qquad\qquad 3x + y = 10$

$4 \stackrel{?}{=} 2 + 2 \qquad\qquad 3(2) + 4 \stackrel{?}{=} 10$

$4 \stackrel{?}{=} 4 \checkmark \qquad\qquad 6 + 4 \stackrel{?}{=} 10$

$\qquad\qquad\qquad\qquad 10 \stackrel{?}{=} 10 \checkmark$

Solve each system by substitution. Check your answer.

1. $\begin{cases} x = y - 1 \\ x + 2y = 8 \end{cases}$ $\underline{(2, 3)}$

2. $\begin{cases} y = x + 2 \\ y = 2x - 5 \end{cases}$ $\underline{(7, 9)}$

3. $\begin{cases} y = x + 5 \\ 3x + y = -11 \end{cases}$ $\underline{(-4, 1)}$

4. $\begin{cases} x = y + 10 \\ x = 2y + 3 \end{cases}$ $\underline{(17, 7)}$

Holt Algebra 1

Holt Algebra 1

Reteach
Solving Systems by Substitution (continued)

You may need to solve one of the equations for a variable before solving with substitution.

Solve $\begin{cases} y - x = 4 \\ 2x + 3y = 27. \end{cases}$

Step 1: Solve the first equation for y.

$$y - x = 4$$
$$\underline{+x \quad +x}$$
$$y = x + 4$$

Step 2: Solve by substitution.

$x + 4$

$2x + 3y = 27$

$2x + 3(x + 4) = 27$ Substitute $x + 4$ for y.

$2x + 3x + 12 = 27$ Distribute.

$5x + 12 = 27$ Combine like terms.

$$\underline{\qquad -12 \quad -12}$$
$$5x = 15$$
$$\frac{5x}{5} = \frac{15}{5}$$
$$x = 3$$

Step 3: Now substitute $x = 3$ back into one of the original equations to find the value of y.

$$y - x = 4$$
$$y - 3 = 4$$
$$\underline{+3 \quad +3}$$
$$y = 7$$

The solution is $(3, 7)$.

Check:

Substitute $(3, 7)$ into both equations.

$$y - x = 4 \qquad\qquad 2x + 3y = 27$$
$$7 - 3 \overset{?}{=} 4 \qquad\quad 2(3) + 3(7) \overset{?}{=} 27$$
$$4 \overset{?}{=} 4 \checkmark \qquad\qquad 6 + 21 \overset{?}{=} 27$$
$$27 \overset{?}{=} 27 \checkmark$$

Solve each system by substitution. Check your answer.

5. $\begin{cases} x - y = -3 \\ 2x + y = 12 \end{cases}$

$(3, 6)$

6. $\begin{cases} y - x = 8 \\ 5x + 2y = 9 \end{cases}$

$(-1, 7)$

15   **Holt Algebra 1**

Challenge
Three Equations in Three Variables

To solve a system with two variables, you must have two equations. To solve a system in three variables, you must have three equations.

Use substitution to find the values of x, y, and z.

1. $\begin{cases} 2x + 4y + 3z = 14 \\ y + 3z = 11 \\ z = 4 \end{cases}$

2. $\begin{cases} 4x + y + 3z = -9 \\ 2y - 5z = 23 \\ z = -3 \end{cases}$

$x = 3;\ y = -1,\ z = 4$ $x = -1;\ y = 4,\ z = -3$

3. $\begin{cases} 2x + 3y + 2z = 18 \\ -4x + y - z = 5 \\ 2z = -10 \end{cases}$

4. $\begin{cases} x + y + z = 15 \\ 2x - y + 3z = -5 \\ -x - y + z = -7 \end{cases}$

[Hint: Solve for z and get two equations to have x and y only.]

$x = 2;\ y = 8,\ z = -5$ $x = -2;\ y = 13,\ z = 4$

16   **Holt Algebra 1**

Problem Solving
Solving Systems by Substitution

Write the correct answer.

1. Maribel has $1.25 in her pocket. The money is in quarters and dimes. There are a total of 8 coins. How many quarters and dimes does Maribel have in her pocket?

3 quarters,

5 dimes

2. Fabulously Fit offers memberships for $35 per month plus a $50 enrollment fee. The Fitness Studio offers memberships for $40 per month plus a $35 enrollment fee. In how many months will the fitness clubs cost the same? What will the cost be?

3 months;

$155

3. Vong grilled 21 burgers at a block party. He grilled the same number of pounds of turkey burgers as hamburgers. Each turkey burger weighed $\frac{1}{4}$ pound and each hamburger weighed $\frac{1}{3}$ pound. How many of each did Vong grill?

12 turkey burgers,

9 beef hamburgers

4. Kate bought 3 used CDs and 1 used DVD at the bookstore. Her friend Joel bought 2 used CDs and 2 used DVDs at the same store. If Kate spent $20 and Joel spent $22, determine the cost of a used CD and a used DVD.

used CD $4.50,

used DVD $6.50

Use the chart below to answer questions 5–8. Select the best answer. The chart compares the quotes that the Masons received from four different flooring contractors to tear out and replace a floor.

5. Which expression shows the total cost if the work is done by Dad's Floors?

A $8 + 150x$ C $150(8x)$
B $150 + 8x$ D $158x$

Contractor	Cost to tear out old floor	Cost of new floor per square foot
Smith & Son	$250	$8.00
V.I.P. Inc.	$350	$7.75
Dad's Floors	$150	$8.00
Floorshop	$300	$8.25

6. How many square feet would the Masons need to have installed to make the total cost of V.I.P. Inc. the same as the total cost of Floorshop?

F 10 sq ft   H 100 sq ft
G 200 sq ft   J 350 sq ft

7. When the total costs of V.I.P. Inc. and Floorshop are the same, what is the total cost?

A $1125.00   C $1950.00
B $1900.00   D $3187.50

8. How many square feet would the Masons need to have installed to make the total cost of Smith & Son the same as the total cost of V.I.P. Inc.?

F 80 sq ft   H 400 sq ft
G 100 sq ft   J 1000 sq ft

17   **Holt Algebra 1**

Reading Strategies
Use a Sequence Chain

Use the sequence chain below to guide you in solving systems of linear equations by the method of substitution.

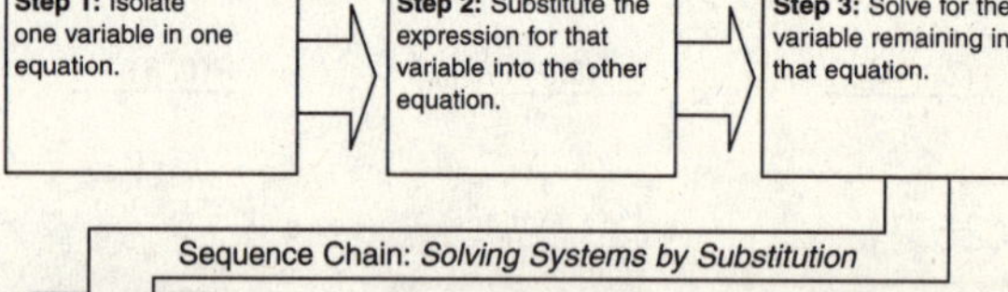

Answer each question.

1. Perform Step 1 for $\begin{cases} x + y = 5 \\ -3x + 5 = \frac{1}{2}y \end{cases}$ by solving the first equation for y.

$$y = 5 - x \text{ or } y = -x + 5$$

2. To solve the system $\begin{cases} y = x + 2 \\ 2x + y = 6 \end{cases}$ by substitution, you can start at Step 2. Explain why.

The first equation is already solved for y.

3. Why does Step 6 specify substituting into *both* equations?

The solution of a system must satisfy both equations.

Solve each system of equations by substitution.

4. $\begin{cases} 2x = 3y \\ y = x - 2 \end{cases}$

$(6, 4)$

5. $\begin{cases} x + y = 2 \\ -x = 2y - 7 \end{cases}$

$(-3, 5)$

18   **Holt Algebra 1**

Holt Algebra 1

Practice A
Solving Systems by Elimination

Fill in the blanks to solve each system by elimination.

1. $\begin{cases} x + 3y = 14 \\ 2x - 3y = -8 \end{cases}$

   Add the equations:
   $x + 3y = 14$
   $+2x - 3y = 28$
   $3x + \underline{0y} = 6$
   $\underline{3x} = 6$
   $\div \underline{3} \div \underline{3}$
   $x = \underline{2}$

   Substitute $\underline{2}$ for x in one of the equations:
   $x + 3y = 14$
   $\underline{2} + 3y = 14$
   $-\underline{2} \quad -\underline{2}$
   $3y = \underline{12}$
   $\div 3 \div 3$
   $y = \underline{4}$
   Solution: ($\underline{2}$, $\underline{4}$)

2. $\begin{cases} 2x + 2y = 4 \\ 3x + 2y = 7 \end{cases}$

   Subtract the equations:
   $2x + 2y = 4$
   $-(3x + 2y = 7)$
   or
   $2x + 2y = 4$
   $-3x - \underline{2y} = \underline{-7}$
   $-x + \underline{0y} = \underline{-3}$
   $-x = \underline{-3}$
   $\div -1 \div -1$
   $x = \underline{3}$

   Substitute $\underline{3}$ for x in one of the equations:
   $3x + 2y = 7$
   $3(\underline{3}) + 2y = 7$
   $\underline{9} + 2y = 7$
   $-\underline{9} \qquad -\underline{9}$
   $2y = \underline{-2}$
   $\div \underline{2} \div \underline{2}$
   $y = \underline{-1}$
   Solution: ($\underline{3}$, $\underline{-1}$)

3. $\begin{cases} 3x + 4y = 26 \\ x - 2y = -8 \end{cases}$

   Multiply the second equation by 2. Then, add the equations:
   $\begin{cases} 3x + 4y = 26 \\ 2(x - 2y = -8) \end{cases}$
   $3x + 4y = 26$
   $+\underline{2}x - \underline{4}y = \underline{-16}$
   $\underline{5}x + 0 = \underline{10}$
   $\underline{5}x = \underline{10}$
   $\div \underline{5} \div \underline{5}$
   $x = \underline{2}$

   Substitute $\underline{2}$ for x in one of the equations:
   $x - 2y = -8$
   $\underline{2} - 2y = -8$
   $-\underline{2} \quad -\underline{2}$
   $-2y = \underline{-10}$
   $\div \underline{-2} \div \underline{-2}$
   $y = \underline{5}$
   Solution: ($\underline{2}$, $\underline{5}$)

Solve each system by elimination.

4. $\begin{cases} 3x - 2y = 1 \\ 2x + 2y = 14 \end{cases}$ $(3, 4)$

5. $\begin{cases} x + y = 4 \\ 3x + y = 16 \end{cases}$ $(6, -2)$

6. $\begin{cases} 3x + 2y = -26 \\ 2x - 6y = -10 \end{cases}$ $(-8, -1)$

7. The sum of two numbers is -1. When twice the first number and four times the second number are added, the sum is -10. What are the two numbers? 3 and -4

    19   **Holt Algebra 1**

Practice B
Solving Systems by Elimination

Follow the steps to solve each system by elimination.

1. $\begin{cases} 2x - 3y = 14 \\ 2x + y = -10 \end{cases}$

   Subtract the second equation:
   $2x - 3y = 14$
   $-(2x + y = -10)$
   $-4y = 24$

   Solve the resulting equation:
   $y = \underline{-6}$

   Use your answer to find the value of x:
   $x = \underline{-2}$
   Solution: ($\underline{-2}$, $\underline{-6}$)

2. $\begin{cases} 3x + y = 17 \\ 4x + 2y = 20 \end{cases}$

   Multiply the first equation by -2. Then, add the equations:
   $-6x - 2y = -34$
   $+ 4x + 2y = 20$
   $-2x = -14$

   Solve the resulting equation:
   $x = \underline{7}$

   Use your answer to find the value of y:
   $y = \underline{-4}$
   Solution: ($\underline{7}$, $\underline{-4}$)

Solve each system by elimination. Check your answer.

3. $\begin{cases} x + 3y = -7 \\ -x + 2y = -8 \end{cases}$ $(2, -3)$

4. $\begin{cases} 3x + y = -26 \\ 2x - y = -19 \end{cases}$ $(-9, 1)$

5. $\begin{cases} x + 3y = -14 \\ 2x - 4y = 32 \end{cases}$ $(4, -6)$

6. $\begin{cases} 4x - y = -5 \\ -2x + 3y = 10 \end{cases}$ $\left(-\dfrac{1}{2}, 3\right)$

7. $\begin{cases} y - 3x = 11 \\ 2y - x = 2 \end{cases}$ $(-4, -1)$

8. $\begin{cases} -10x + y = 0 \\ 5x + 3y = -7 \end{cases}$ $\left(-\dfrac{1}{5}, -2\right)$

Solve.

9. Brianna's family spent $134 on 2 adult tickets and 3 youth tickets at an amusement park. Max's family spent $146 on 3 adult tickets and 2 youth tickets. What is the price of a youth ticket? $22

10. Carl bought 19 apples of 2 different varieties to make a pie. The total cost of the apples was $5.10. Granny Smith apples cost $0.25 each and Gala apples cost $0.30 each. How many of each type of apple did Carl buy?   7 Gala apples; 12 Granny Smith apples

    20   **Holt Algebra 1**

Practice C
Solving Systems by Elimination

Solve each system by elimination.

1. $\begin{cases} x + y = 2 \\ 2x - y = 7 \end{cases}$ $(3, -1)$

2. $\begin{cases} 3x - 2y = -2 \\ 3x + y = 10 \end{cases}$ $(2, 4)$

3. $\begin{cases} x + y = -7 \\ x - y = 5 \end{cases}$ $(-1, -6)$

4. $\begin{cases} -3x - 4y = -2 \\ 6x + 4y = 3 \end{cases}$ $\left(\dfrac{1}{3}, \dfrac{1}{4}\right)$

5. $\begin{cases} 2x - 2y = 14 \\ x + 4y = -13 \end{cases}$ $(3, -4)$

6. $\begin{cases} y - x = 17 \\ 2y + 3x = -11 \end{cases}$ $(-9, 8)$

7. $\begin{cases} x + 6y = 1 \\ 2x - 3y = 32 \end{cases}$ $(13, -2)$

8. $\begin{cases} -\dfrac{1}{2}x + y = 4 \\ \dfrac{1}{3}x - y = -3 \end{cases}$ $(-6, 1)$

9. $\begin{cases} 3x + y = -15 \\ 2x - 3y = 23 \end{cases}$ $(-2, -9)$

10. $\begin{cases} 5x - 2y = -48 \\ 2x + 3y = -23 \end{cases}$ $(-10, -1)$

11. $\begin{cases} 4x - 3y = -9 \\ 5x - y = 8 \end{cases}$ $(3, 7)$

12. $\begin{cases} 3x - 3y = -1 \\ 12x - 2y = 16 \end{cases}$ $\left(\dfrac{5}{3}, 2\right)$

13. At a bakery, Riley bought 3 bagels and 2 muffins for $7.25. Karen bought 5 bagels and 4 muffins for $13.25. What is the cost of each item?   bagel: $1.25; muffin: $1.75

14. A chemist has a beaker of a 3% acid solution and a beaker of a 7% acid solution. He needs to make 75 mL of a 4% acid solution.

   a. Complete the table.

	3% solution	+	7% solution	=	4% solution
Amount of Solution (mL)	x	+	y	=	75
Amount of Acid (mL)	0.03x	+	0.07y	=	0.04(75)

   b. Use the information in the table to write a system of linear equations.
   $\begin{cases} x + y = 75 \\ 0.03x + 0.07y = 3 \end{cases}$

   c. Solve the system of equations to find how much he will use from each beaker.
   56.25 mL of the 3% solution; 18.75 mL of the 7% solution

    21   **Holt Algebra 1**

Reteach
Solving Systems by Elimination

Elimination can be used to solve a system of equations by adding terms vertically. This will cause one of the variables to be eliminated. It may be necessary to multiply one or both equations by some number to use this method.

I. Elimination may require no change to either equation.

$\begin{cases} 3x + y = 6 \\ 5x - y = 10 \end{cases}$ Adding vertically will eliminate y. $\begin{array}{l} 3x + y = 6 \\ 5x - y = 10 \\ \hline 8x + 0 = 16 \end{array}$

II. Elimination may require multiplying one equation by an appropriate number.

$\begin{cases} 2x + 5y = 9 \\ x - 3y = 10 \end{cases}$ Multiply bottom equation by -2. $\begin{array}{l} 2x + 5y = 9 \\ -2(x - 3y) = -2(10) \end{array}$ $\begin{array}{l} 2x + 5y = 9 \\ -2x + 6y = -20 \\ \hline 0 + 11y = -11 \end{array}$

III. Elimination may require multiplying both equations by different numbers.

$\begin{cases} 5x + 3y = 2 \\ 4x + 2y = 10 \end{cases}$ Multiply the top by -2 and the bottom by 3. $\begin{array}{l} -2(5x + 3y = 2) \\ 3(4x + 2y = 10) \end{array}$ $\begin{array}{l} -10x + -6y = -4 \\ 12x + 6y = 30 \\ \hline 2x + 0 = 26 \end{array}$

Solve each system by elimination.

1. $\begin{cases} 2x - y = 20 \\ 3x + 2y = -19 \end{cases}$ $(3, -14)$

2. $\begin{cases} 3x + 2y = 10 \\ 3x - 2y = 14 \end{cases}$ $(4, -1)$

3. $\begin{cases} x + y = 12 \\ 2x + y = 6 \end{cases}$ $(-6, 18)$

4. $\begin{cases} 3x - y = 2 \\ -8x + 2y = 4 \end{cases}$ $(-4, -14)$

    22   **Holt Algebra 1**

Reteach
Solving Systems by Elimination (continued)

A system of equations can be solved by graphing, substitution, or elimination.

- Use graphing if both equations are solved for y, or if you want an estimate of the solution.
- Use substitution if either equation is solved for a variable, or has a variable with a coefficient of 1 or -1.
- Use elimination if both equations have the same variable with the same or opposite coefficients.

It may be necessary to manipulate your equations to get them in any of the three forms above.

Solve $\begin{cases} y = 3 - x \\ 2x - y = 6 \end{cases}$

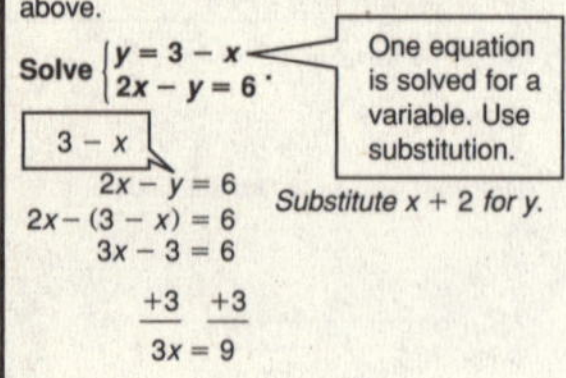

$3 - x$

$2x - y = 6$

$2x - (3 - x) = 6$ Substitute $x + 2$ for y.

$3x - 3 = 6$

$\ +3\ \ +3$

$3x = 9$

$x = 3$

Substitute $x = 3$ into one of the original equations to find the value of y.

$y = 3 - x$

$y = 3 - 3$

$y = 0$ The solution is $(3, 0)$.

Solve $\begin{cases} -2x - y = -5 \\ 3x + y = -1 \end{cases}$

The equations have the same variable with opposite coefficients. Use elimination.

$-2x - y = -5$

$\underline{3x + y = -1}$

$x + 0 = -6$

$x = -6$

Substitute $x = -6$ into one of the original equations to find the value of y.

$3x + y = -1$

$3(-6) + y = -1$

$-18 + y = -1$

$\ +18\ \ \ +18$

$y = 17$

The solution is $(-6, 17)$.

Solve each system by any method.

5. $\begin{cases} y = x + 3 \\ -2x + y = -4 \end{cases}$

6. $\begin{cases} 4x + y = 10 \\ -2x - y = 4 \end{cases}$

7. $\begin{cases} 2x + y = 8 \\ 3x + 5y = 5 \end{cases}$

$(7, 10)$ $(7, -18)$ $(5, -2)$

23   **Holt Algebra 1**

Challenge
Solving Systems by Elimination

The elimination method can also be used for a system of three equations in three unknowns.

Three camp leaders purchased equipment for a camping trip.
Max bought 10 sleeping bags, 2 tents, and 1 can of bug repellant for $885.
Carlos bought 5 sleeping bags, 4 tents, and 1 can of bug repellant for $865.
Amy bought 9 sleeping bags, 6 tents, and 6 cans of bug repellant for $1410.
If they made their purchases at the same store, how much did each item cost?

1. Write the 3 equations: $\begin{cases} 10x + 2y + z = 885 \\ 5x + 4y + z = 865 \\ 9x + 6y + 6z = 1410 \end{cases}$

2. Subtract the second equation from the first. $5x - 2y = 20$

3. Multiply the second equation by -6 and add the second and third equations. $-21x - 18y = -3780$

4. The equations in steps 2 and 3 form a linear system in two variables. Solve this system for x. $x = 60$

5. Substitute the value of x into the first two equations. Write the resulting system. $\begin{cases} 2y + z = 285 \\ 4y + z = 565 \end{cases}$

6. Solve the system in problem 5 for y and z. $y = 140, z = 5$

7. Write the cost of each item.   sleeping bags: $60; tents: $140; bug repellant: $5

24   **Holt Algebra 1**

Problem Solving
Solving Systems by Elimination

Write the correct answer.

1. Mr. Nguyen bought a package of 3 chicken legs and a package of 7 chicken wings. Ms. Dawes bought a package of 3 chicken legs and a package of 6 chicken wings. Mr. Nguyen bought 45 ounces of chicken. Ms. Dawes bought 42 ounces of chicken. How much did each chicken leg and each chicken wing weigh?

chicken leg 8 oz.,

chicken wing 3 oz.

2. Jayce bought 2 bath towels and returned 3 hand towels. His sister Jayna bought 3 bath towels and 3 hand towels. Jayce's bill was $5. Jayna's bill was $45. What are the prices of a bath towel and a hand towel?

bath towel $10,

hand towel $5

3. The Lees spent $31 on movie tickets for 2 adults and 3 children. The Macias spent $26 on movie tickets for 2 adults and 2 children. What are the prices for adult and child movie tickets?

adult ticket $8,

child ticket $5

4. Last month Stephanie spent $57 on 4 allergy shots and 1 office visit. This month she spent $9 after 1 office visit and a refund for 2 allergy shots from her insurance company. How much does an office visit cost? an allergy shot?

office visit $25,

allergy shot $8

Use the chart below to answer questions 5–6. Select the best answer.
The chart shows the price per pound for dried fruit.

Dried Fruit Price List			
Pineapple	Apple	Mango	Papaya
$7.50/lb	$7.00/lb	$8.00/lb	$7.25/lb

5. A customer bought 5 pounds of mango and papaya for $37.75. How many pounds of each fruit did the customer buy?
   A) 2 lbs mango and 3 lbs papaya
   B) 3 lbs mango and 2 lbs papaya
   C) 1 lb mango and 4 lbs papaya
   D) 4 lbs mango and 1 lb papaya

6. A store employee made two gift baskets of dried fruit, each costing $100. The first basket had 12 pounds of fruit x and 2 pounds of fruit y. The second basket had 4 pounds of fruit x and 9 pounds of fruit y. Which two fruits did the employee use in the baskets?
   F) pineapple and apple
   G) apple and mango
   H) mango and papaya
   J) papaya and pineapple

25   **Holt Algebra 1**

Reading Strategies
Connecting Concepts

When solving systems of linear equations using elimination, you will sometimes need to multiply one or both equations by a factor in order to get the same coefficients for a variable. This process is very similar to getting a common denominator for fractions. Look at the example below.

$\begin{cases} 6x - 5y = 16 \\ 4x - 3y = 12 \end{cases}$ → To eliminate the x-terms, you need to get the same or opposite coefficients for x in both equations.

Add: $\dfrac{5}{6} + \dfrac{1}{4} = ?$ → Think about finding a common denominator for 4 and 6.

4: 4, 8, 12, 16, 20
6: 6, 12, 18, 24, 30 → Find the least common multiple (LCM) of 4 and 6 by listing their multiples in order. The LCM is 12.

$\begin{cases} 2(6x - 5y = 16) \\ 3(4x - 3y = 12) \end{cases}$ → Determine what you have to multiple 4 and 6 by to get 12. Multiply each equation by the appropriate number.

$\begin{cases} 12x - 10y = 32 \\ 12x - 9y = 36 \end{cases}$ → Now either add or subtract the equations. In this case, you will subtract the equations.

1. Describe how you would get common y-coefficients (instead of x-) in the example above.

Multiply the first equation by 3 and the second equation by 5 to get common coefficients of -15.

2. Show how to get a set of common y-coefficients for the system $\begin{cases} 9x - 10y = 7 \\ 5x + 8y = 31 \end{cases}$

$\begin{cases} 4(9x - 10y = 7) \\ 5(5x + 8y = 31) \end{cases} \Rightarrow \begin{cases} 36x - 40y = 28 \\ 25x + 40y = 155 \end{cases}$

Solve each system of equations by elimination.

3. $\begin{cases} 9x - 2y = 15 \\ 4x + 3y = -5 \end{cases}$

4. $\begin{cases} 2x - 3y = 50 \\ 7x + 8y = -10 \end{cases}$

$(1, -3)$ $(10, -10)$

26   **Holt Algebra 1**

Holt Algebra 1

Practice A
Solving Special Systems

Solve each system of linear equations. Tell whether the system has no solution or infinitely many solutions.

1. $\begin{cases} 2x + y = 1 \\ 2x + y = -3 \end{cases}$

   no solution

2. $\begin{cases} y = 5x + 2 \\ y - 5x = 2 \end{cases}$

   infinitely many solutions

3. $\begin{cases} y - 3x + 2 = 0 \\ 2 = -y + 3x \end{cases}$

   infinitely many solutions

4. $\begin{cases} x + y = 4 \\ y - 4 = 1 - x \end{cases}$

   no solution

Give the number of solutions to each system. Then classify the system as "consistent, independent", "consistent, dependent", or "inconsistent".

5. $\begin{cases} y = 2(x + 1) \\ y - 2x = 2 \end{cases}$

   infinitely many solutions;
   consistent, dependent

6. $\begin{cases} y - 4x + 5 = 0 \\ 4x = y - 1 \end{cases}$

   no solution;
   inconsistent

7. Marquis opens a savings account with $60 and adds $20 each month. His brother Jibran adds $20 each month to the savings account that his grandmother opened with $60. If the brothers continue to make deposits to their savings accounts at the same rate, when will they have the same amount of money? Explain.

   They will always have the same amount of money.
   The graphs of these equations are the same line.

27
Holt Algebra 1

Practice B
Solving Special Systems

Solve each system of linear equations.

1. $\begin{cases} y = 2x - 3 \\ y - 2x = -3 \end{cases}$

   infinitely many solutions

2. $\begin{cases} 3x + y = 4 \\ -3x = y - 7 \end{cases}$

   no solution

3. $\begin{cases} y = -4x + 1 \\ 4x = -y - 6 \end{cases}$

   no solution

4. $\begin{cases} y - x + 3 = 0 \\ x = y + 3 \end{cases}$

   infinitely many solutions

Classify each system. Give the number of solutions.

5. $\begin{cases} y = 3(x - 1) \\ -y + 3x = 3 \end{cases}$

   consistent, dependent;
   infinitely many solutions

6. $\begin{cases} y - 2x = 5 \\ x = y - 3 \end{cases}$

   consistent, independent;
   one solution

7. Sabina and Lou are reading the same book. Sabina reads 12 pages a day. She had read 36 pages when Lou started the book, and Lou reads at a pace of 15 pages per day. If their reading rates continue, will Sabina and Lou ever be reading the same page on the same day? Explain.

   Yes. The graphs of the two
   equations have different
   slopes. They will
   intersect.

8. Brandon started jogging at 4 miles per hour. After he jogged 1 mile, his friend Anton started jogging along the same path at a pace of 4 miles per hour. If they continue to jog at the same rate, will Anton ever catch up with Brandon? Explain.

   No. The graphs of the two
   equations are parallel
   lines. They will never
   intersect.

28
Holt Algebra 1

Practice C
Solving Special Systems

Solve each system of linear equations.

1. $\begin{cases} y + 2x + 4 = 0 \\ 2x = -y - 4 \end{cases}$

   infinitely many solutions

2. $\begin{cases} 5x + y = 8 \\ -5x = 8 + y \end{cases}$

   no solution

3. $\begin{cases} 2x - y = 4 \\ 1 = y - 2x + 5 \end{cases}$

   infinitely many solutions

4. $\begin{cases} y = -x - 6 \\ y - 2x = -3x + 6 \end{cases}$

   no solution

Classify each system. Give the number of solutions.

5. $\begin{cases} y + 2(x - 3) = 0 \\ 2x = -y - 3 \end{cases}$

   inconsistent;
   no solution

6. $\begin{cases} y + 3x = -1 \\ x = y + 3x - 1 \end{cases}$

   consistent, independent;
   one solution

7. At a factory, Jin assembles 12 parts each minute. He has assembled 156 parts when Summer starts on the line, assembling at a pace of 15 parts per minute. If their assembly rates continue, will Summer ever catch up to Jin? Explain.

   Yes. The graphs of the two
   equations have different
   slopes. They will
   intersect.

8. Kat is comparing monthly sales at her bookstore with those of her competitor, Gill. If the sales rates continue, will Kat's book sales ever catch up with her competitor's? Explain.

Kat's sales	112	118	124	130
Gill's sales	138	144	150	156

   No. The graphs of the two
   equations are parallel
   lines. They will never
   intersect.

29
Holt Algebra 1

Reteach
Solving Special Systems

When solving equations in one variable, it is possible to have one solution, no solutions, or infinitely many solutions. The same results can occur when graphing systems of equations.

Solve $\begin{cases} 4x + 2y = 2 \\ 2x + y = 4 \end{cases}$

Multiplying the second equation by -2 will eliminate the x-terms.

$4x + 2y = 2$
$-2(2x + y = 4)$ $\rightarrow$

$4x + 2y = 2$
$\underline{-4x - 2y = -8}$
$0 + 0 = -6$
$0 = -6$ ✗

The equation is a contradiction. **There is no solution.**

Solve $\begin{cases} y = 4 - 3x \\ 3x + y = 4 \end{cases}$

Because the first equation is solved for a variable, use substitution.

$3x + y = 4$
$3x + (4 - 3x) = 4$ *Substitute* $4 - 3x$ *for* y
$0 + 4 = 4$
$4 = 4$ ✓

The equation is true for all values of x and y. **There are infinitely many solutions.**

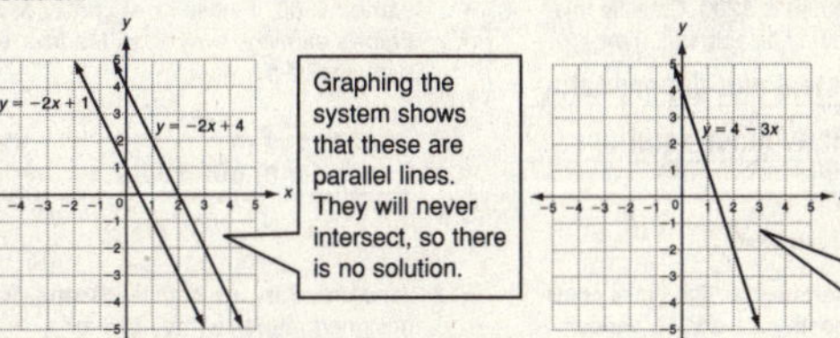

Solve each system of linear equations algebraically.

1. $\begin{cases} y = 3x \\ 2y = 6x \end{cases}$

   infinitely many solutions

2. $\begin{cases} y = 2x + 5 \\ y - 2x = 1 \end{cases}$

   no solution

3. $\begin{cases} 3x - 2y = 9 \\ -6x + 4y = 1 \end{cases}$

   no solution

30
Holt Algebra 1

Holt Algebra 1

Reteach
Solving Special Systems (continued)

A system of linear equations can be classified in three ways.

I. Consistent and independent one solution different slopes	Example: $\begin{cases} y = x + 3 \\ y = -x + 6 \end{cases}$	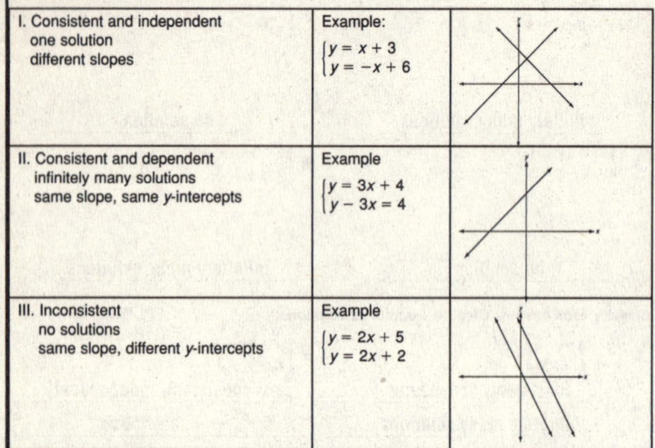
II. Consistent and dependent infinitely many solutions same slope, same y-intercepts	Example $\begin{cases} y = 3x + 4 \\ y - 3x = 4 \end{cases}$	
III. Inconsistent no solutions same slope, different y-intercepts	Example $\begin{cases} y = 2x + 5 \\ y = 2x + 2 \end{cases}$	

Classify each system below by comparing the slopes and y-intercepts. Then give the number of solutions.

4. $\begin{cases} y = -3x - 2 \\ y = -3x - 4 \end{cases}$

__inconsistent;__

__no solutions__

5. $\begin{cases} y = 2x + 5 \\ y = 5 + 2x \end{cases}$

__consistent and dependent;__

__infinitely many solutions__

6. $\begin{cases} y = -4x + 3 \\ y = 2x + 7 \end{cases}$

__consistent and independent;__

__one solution__

Classify each system and give the number of solutions. If there is one solution, provide it.

7. $\begin{cases} y = 2x + 8 \\ y - 4x = 8 \end{cases}$

__consistent and independent;__

__one solution; (0, 8)__

8. $\begin{cases} y + 3x - 2 = 0 \\ 9x + 3y = 6 \end{cases}$

__consistent and dependent;__

__infinitely many solutions__

31
Holt Algebra 1

Challenge
Pick a Path

The puzzle below contains 23 linear equations. You goal is to find paths through the puzzle such that each pair of equations along your path forms a special type of system.

You must begin in the **Start** square, and end in the **Finish** square. Moves can be horizontal, vertical, or diagonal between adjacent squares. Each individual path cannot cross itself, but your answers to 1 and 2 might cross.

	1	2	3	4
Start →	$y = x + 2$	$3x = 3y - 6$	$3x + y = -1$	$8y = 8x + 16$
5 $y = 2x + 3$	**6** $y = 3x - 1$	**7** $y - x = 2$	**8** $\frac{1}{3}x + \frac{1}{3}y = 1$	**9** $3x = y + 1$
10 $2y = 6x + 6$	**11** $2x - y = -3$	**12** $2x + 2y = 4$	**13** $2y - 4 = 2x$	**14** $x - y = -2$
15 $\frac{1}{2}y = \frac{3}{2}x + \frac{1}{2}$	**16** $3y = 2x + 1$	**17** $3x - y = 4$	**18** $\frac{1}{6}y = \frac{1}{3}x + \frac{1}{2}$	**19** $\frac{1}{2}y = \frac{1}{2}x + 1$
20 $10x + 15y = 5$	**21** $-12x + 4y = 8$	**22** $\frac{1}{2}y = \frac{1}{3}x + \frac{1}{6}$	**23** $x = \frac{1}{3}y + 1$	**→ Finish**

Find a path through the puzzle such that each pair of equations forms...

1. ...a consistent and dependent system.

Start − 1 − 2 − 7 − 13 − 14 − 19 − Finish

2. ...an inconsistent system.

Start − 6 − 10 − 15 − 21 − 17 − 23 − Finish

32
Holt Algebra 1

Problem Solving
Solving Special Systems

Write the correct answer.

1. Tyra and Charmian are training for a bike race. Tyra has logged 256 miles so far and rides 48 miles per week. Charmian has logged 125 miles so far and rides 48 miles per week. If these rates continue, will Tyra's distance ever equal Charmian's distance? Explain.

__No; the graphs are parallel__

__lines so there is no solution.__

2. Metroplexpress and Local Express are courier companies. Metroplexpress charges $15 to pick up a package and $0.50 per mile. Local Express charges $10 to pick up a package and $0.55 per mile. Classify this system and find its solution, if any.

__consistent and independent;__

__100 mi and $65__

3. The Singhs start savings accounts for their twin boys. The accounts earn 5% annual interest. The initial deposit in each account is $200. Classify this system and find its solution, if any.

__consistent and dependent;__

__infinitely many solutions__

4. Frank earns $8 per hour. Madison earns $7.50 per hour. Frank started working after Madison had already earned $300. If these rates continue, will Frank's earnings ever equal Madison's earnings? If so, when?

__Yes;__

__at 600 hours.__

Select the best answer.

5. A studio apartment at The Oaks costs $400 per month plus a $350 deposit. A studio apartment at Crossroads costs $400 per month plus a $300 deposit. How many solutions does this system have?

(A) no solutions
B 1 solution
C 2 solutions
D an infinite number of solutions

6. Jane and Gary are both landscape designers. Jane charges $75 for a consultation plus $25 per hour. Gary charges $50 for a consultation plus $30 per hour. For how many hours will Jane's charges equal Gary's charges?

F never
G after 2 hours
(H) after 5 hours
J always

7. A tank filled with 75 liters of water loses 0.5 liter of water per hour. A tank filled with 50 liters of water loses 0.1 liter of water per hour. How would this system be classified?

A inconsistent
B dependent
(C) consistent and independent
D consistent and dependent

8. Simon is 3 years older than Renata. Five years ago, Renata was half as old as Simon is now. How old are Simon and Renata now?

F Simon is 13 and Renata is 10.
G Simon is 15 and Renata is 12.
H Simon is 16 and Renata is 8.
(J) Simon is 16 and Renata is 13.

33
Holt Algebra 1

Reading Strategies
Use a Table

The table below can help you answer questions about linear systems.

Classification	Number of Solutions	Similarities and Differences in $y = mx + b$	Description of Graphed Lines	Result of Solving with Algebra
Consistent, Independent	1	different slopes (m)	intersecting	values for x and y ex. $x = 3$, $y = -5$
Consistent, Dependent	infinitely many	same slope (m) same y-int. (b)	coincident	identity statement ex. $4 = 4$
Inconsistent	0	same slope (m) different y-int. (b)	parallel	false statement ex. $-2 = 3$

1. Mary Kate solved a system by elimination as shown below. Classify the system.

$\begin{cases} 2x + 3y = 5 \\ 2x + 3y = 7 \end{cases}$ → $\begin{aligned} 2x + 3y &= 5 \\ -(2x + 3y &= 7) \\ \hline 0 &= -2 \end{aligned}$

__inconsistent__

2. Raul solved a system of equations by substitution. He ended up with an identity statement. How many solutions does his system have?

__infinitely many__

3. How many solutions does a system have if, when graphed, the lines are the same line?

__infinitely many__

4. Two equations in a system have the same slope. Which classification can NOT describe the system?

__consistent, independent__

5. The graph of a system consists of two intersecting lines. How many solutions does the system have?

__one__

Classify each system, give the number of solutions, and describe its graph.

6. $\begin{cases} y = -3x \\ y = -3x + 2 \end{cases}$

__inconsistent;__

__0;__

__parallel lines__

7. $\begin{cases} y = -x + 4 \\ x + y = 4 \end{cases}$

__consistent, dependent;__

__infinitely many;__

__coincident lines__

8. $\begin{cases} y = 2x - 1 \\ x = y \end{cases}$

__consistent, independent;__

__1;__

__intersecting lines__

34
Holt Algebra 1

Holt Algebra 1

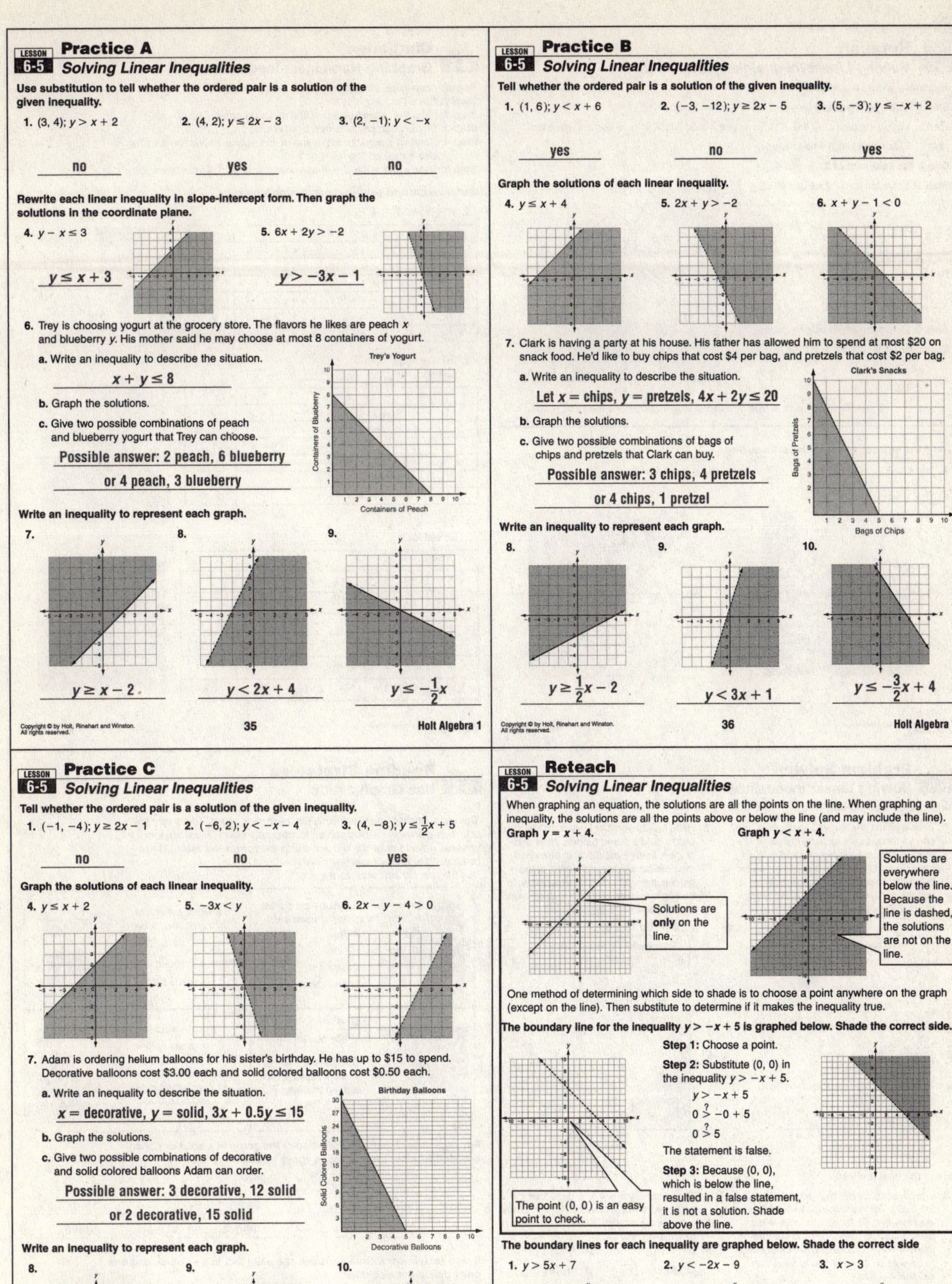

Practice A
6-5 Solving Linear Inequalities

Use substitution to tell whether the ordered pair is a solution of the given inequality.

1. $(3, 4); y > x + 2$

2. $(4, 2); y \le 2x - 3$

3. $(2, -1); y < -x$

no   yes   no

Rewrite each linear inequality in slope-intercept form. Then graph the solutions in the coordinate plane.

4. $y - x \le 3$

$y \le x + 3$

5. $6x + 2y > -2$

$y > -3x - 1$

6. Trey is choosing yogurt at the grocery store. The flavors he likes are peach x and blueberry y. His mother said he may choose at most 8 containers of yogurt.

a. Write an inequality to describe the situation.

$x + y \le 8$

b. Graph the solutions.

c. Give two possible combinations of peach and blueberry yogurt that Trey can choose.

Possible answer: 2 peach, 6 blueberry or 4 peach, 3 blueberry

Trey's Yogurt

Write an inequality to represent each graph.

7. $y \ge x - 2$

8. $y < 2x + 4$

9. $y \le -\dfrac{1}{2}x$

35   **Holt Algebra 1**

Practice B
6-5 Solving Linear Inequalities

Tell whether the ordered pair is a solution of the given inequality.

1. $(1, 6); y < x + 6$

2. $(-3, -12); y \ge 2x - 5$

3. $(5, -3); y \le -x + 2$

yes   no   yes

Graph the solutions of each linear inequality.

4. $y \le x + 4$

5. $2x + y > -2$

6. $x + y - 1 < 0$

7. Clark is having a party at his house. His father has allowed him to spend at most $20 on snack food. He'd like to buy chips that cost $4 per bag, and pretzels that cost $2 per bag.

a. Write an inequality to describe the situation.

Let x = chips, y = pretzels, $4x + 2y \le 20$

b. Graph the solutions.

c. Give two possible combinations of bags of chips and pretzels that Clark can buy.

Possible answer: 3 chips, 4 pretzels or 4 chips, 1 pretzel

Clark's Snacks

Write an inequality to represent each graph.

8. $y \ge \dfrac{1}{2}x - 2$

9. $y < 3x + 1$

10. $y \le -\dfrac{3}{2}x + 4$

36   **Holt Algebra 1**

Practice C
6-5 Solving Linear Inequalities

Tell whether the ordered pair is a solution of the given inequality.

1. $(-1, -4); y \ge 2x - 1$

2. $(-6, 2); y < -x - 4$

3. $(4, -8); y \le \dfrac{1}{2}x + 5$

no   no   yes

Graph the solutions of each linear inequality.

4. $y \le x + 2$

5. $-3x < y$

6. $2x - y - 4 > 0$

7. Adam is ordering helium balloons for his sister's birthday. He has up to $15 to spend. Decorative balloons cost $3.00 each and solid colored balloons cost $0.50 each.

a. Write an inequality to describe the situation.

x = decorative, y = solid, $3x + 0.5y \le 15$

b. Graph the solutions.

c. Give two possible combinations of decorative and solid colored balloons Adam can order.

Possible answer: 3 decorative, 12 solid or 2 decorative, 15 solid

Birthday Balloons

Write an inequality to represent each graph.

8. $y \ge \dfrac{1}{3}x - 4$

9. $y < 2x$

10. $y \le 3$

37   **Holt Algebra 1**

Reteach
6-5 Solving Linear Inequalities

When graphing an equation, the solutions are all the points on the line. When graphing an inequality, the solutions are all the points above or below the line (and may include the line).

Graph $y = x + 4$.

Graph $y < x + 4$.

Solutions are **only** on the line.

Solutions are everywhere below the line. Because the line is dashed, the solutions are not on the line.

One method of determining which side to shade is to choose a point anywhere on the graph (except on the line). Then substitute to determine if it makes the inequality true.

The boundary line for the inequality $y > -x + 5$ is graphed below. Shade the correct side.

The point $(0, 0)$ is an easy point to check.

Step 1: Choose a point.

Step 2: Substitute $(0, 0)$ in the inequality $y > -x + 5$.

$y > -x + 5$

$0 \overset{?}{>} -0 + 5$

$0 \overset{?}{>} 5$

The statement is false.

Step 3: Because $(0, 0)$, which is below the line, resulted in a false statement, it is not a solution. Shade above the line.

The boundary lines for each inequality are graphed below. Shade the correct side

1. $y > 5x + 7$

2. $y < -2x - 9$

3. $x > 3$

38   **Holt Algebra 1**

     Holt Algebra 1

Reteach
Solving Linear Inequalities (continued)

To graph a linear inequality:

Step 1: Solve the inequality for y.

Step 2: Graph the boundary line. If $\leq$, or $\geq$ use a solid line. If $<$ or $>$ use a dashed line.

Step 3: Determine which side to shade.

Graph the solutions of $2x + y \leq 4$.

Step 1: Solve for y.

$$2x + y \leq 4$$
$$\underline{-2x \quad -2x}$$
$$y \leq -2x + 4$$

Step 2: Graph the boundary line.

Use a solid line for $\leq$.

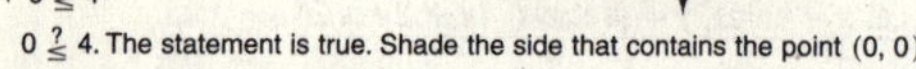

Step 3: Determine which side to shade.

Substitute $(0, 0)$ into $2x + y \leq 4$.

$$2x + y \leq 4$$
$$2(0) + 0 \overset{?}{\leq} 4$$
$$0 \overset{?}{\leq} 4.$$ The statement is true. Shade the side that contains the point $(0, 0)$.

Graph the solutions of each linear inequality.

4. $y - x < 3$

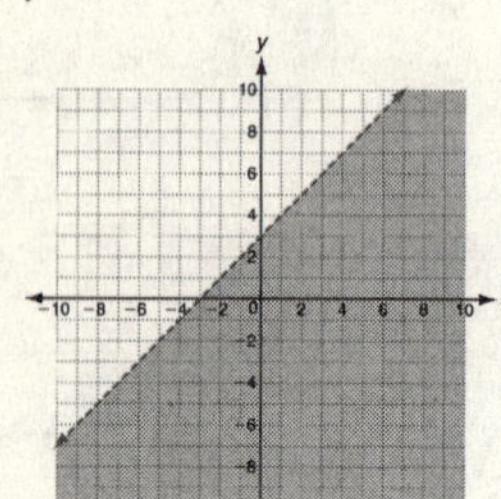

5. $x + y + 2 \geq 0$

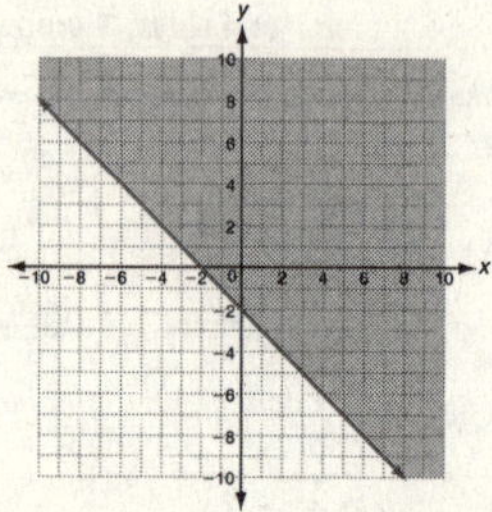

39

Challenge
Graphing Non-Linear Inequalities

To graph non-linear inequalities, follow the steps below.

Step 1: Solve the inequality for y.
Step 2: Write the related equation and generate ordered pairs.
Step 3: Plot enough points to see a pattern.
Step 4: Connect the points with a line or curve. Use a solid line for $\leq$ or $\geq$. Use a dashed line for $<$ or $>$.
Step 5: Shade above the line or curve for $y >$ or $y \geq$. Shade below for $y <$ or $y \leq$.

Generate ordered pairs and graph each inequality.

1. $y > (x + 3)^2 - 4$

x	Think: $y = (x + 3)^2 - 4$	(x, y)
-6	$y = (-6 + 3)^2 - 4 = (-3)^2 - 4 = 9 - 4 = 5$	$(-6, 5)$
-5	$y = (-5 + 3)^2 - 4 = (-2)^2 - 4 = 4 - 4 = 0$	$(-5, 0)$
-4	$y = (-4 + 3)^2 - 4 = (-1)^2 - 4 = 1 - 4 = -3$	$(-4, -3)$
-3	$y = (-3 + 3)^2 - 4 = (0)^2 - 4 = 0 - 4 = -4$	$(-3, -4)$
-2	$y = (-2 + 3)^2 - 4 = (1)^2 - 4 = 1 - 4 = -3$	$(-2, -3)$
-1	$y = (-1 + 3)^2 - 4 = (2)^2 - 4 = 4 - 4 = 0$	$(-1, 0)$

2. $|x| + y \leq 5$
   Solved for y: $\quad y \leq -|x| + 5$

x	Think: $y = -|x| + 5$	(x, y)
-3	$y = -|-3| + 5 = -(3) + 5 = -3 + 5 = 2$	$(-3, 2)$
-2	$y = -|-2| + 5 = -(2) + 5 = -2 + 5 = 3$	$(-2, 3)$
-1	$y = -|-1| + 5 = -(1) + 5 = -1 + 5 = 4$	$(-1, 4)$
0	$y = -|0| + 5 = 0 + 5 = 5$	$(0, 5)$
1	$y = -|1| + 5 = -1 + 5 = 4$	$(1, 4)$
2	$y = -|2| + 5 = -2 + 5 = 3$	$(2, 3)$

3. $2x^2 - 4y > 8$
   Solved for y: $\quad y < \frac{1}{2}x^2 - 2$

x	Think: $y = \frac{1}{2}x^2 - 2$	(x, y)
-4	$y = \frac{1}{2}(-4)^2 - 2 = \frac{1}{2}(16) - 2 = 8 - 2 = 6$	$(-4, 6)$
-2	$y = \frac{1}{2}(-2)^2 - 2 = \frac{1}{2}(4) - 2 - 2 - 2 = 0$	$(-2, 0)$
0	$y = \frac{1}{2}(0)^2 - 2 = \frac{1}{2}(0) - 2 = 0 - 2 = -2$	$(0, -2)$
1	$y = \frac{1}{2}(1)^2 - 2 = \frac{1}{2}(1) - 2 = \frac{1}{2} - 2 = -1\frac{1}{2}$	$\left(1, -1\frac{1}{2}\right)$
2	$y = \frac{1}{2}(2)^2 - 2 = \frac{1}{2}(4) - 2 = 2 - 2 = 0$	$(2, 0)$
4	$y = \frac{1}{2}(4)^2 - 2 = \frac{1}{2}(16) - 2 = 8 - 2 = 6$	$(4, 6)$

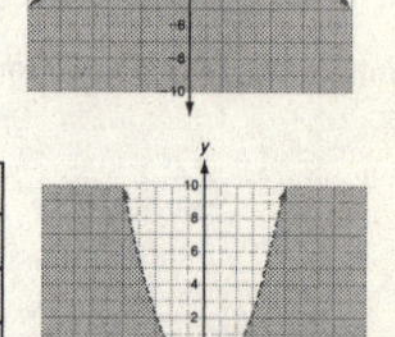

40

Problem Solving
Solving Linear Inequalities

Write the correct answer.

1. Shania would like to give $5 gift cards and $4 teddy bears as party favors. Sixteen people have been invited to the party. Shania has $100 to spend on party favors. Write and graph an inequality to find the number of gift cards x and teddy bears y Shania could purchase.

$$5x + 4y \leq 100$$

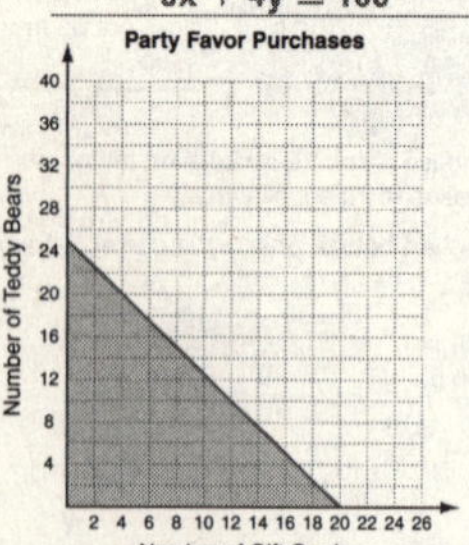

2. Hank has 20 yards of lumber that he can use to build a raised garden. Write and graph a linear inequality that describes the possible lengths and widths of the garden. If Hank wants the dimensions to be whole numbers only, what dimensions would produce the largest area?

$$2x + 2y \geq 20; \text{ 5 yd by 5 yd}$$

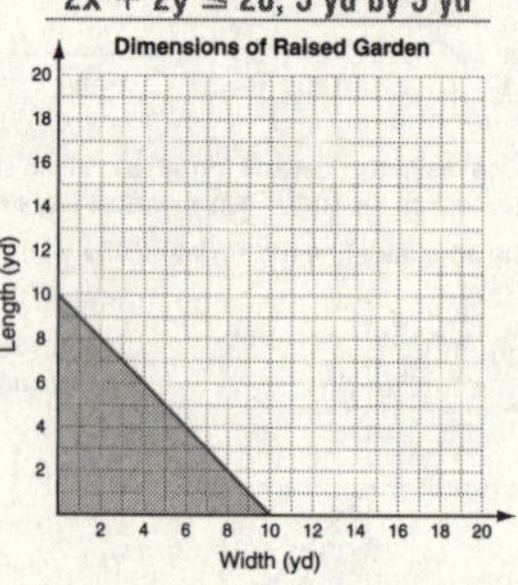

Select the best answer.

3. The royalties for the high school play are $250. Tickets to the play cost $5 for students and $8 for nonstudents. What linear inequality describes the number of student and nonstudent tickets that need to be sold so that the drama class can pay the royalties?
   A $5x + 8y \geq 250$ **C** $5xy + 8 < 250$
   B $5x + 8y > 250$ **D** $5xy + 8 \geq 250$

4. The inequality $x + y \leq 8$ describes the amounts of two juices Annette combines to make a smoothie. Which is a solution to the inequality?
   F $(3, 6)$ **H** $(7, 2)$
   G $(6, 1)$ **J** $(0, 10)$

5. A baker is making chocolate and lemon pound cakes. He can make at most 12 cakes at one time. Which inequality describes the situation?
   A $x + y > 12$ **C** $x + y \leq 12$
   B $x + y \geq 12$ **D** $x + y < 12$

6. Erasmus is the master gardener for a university. He wants to plant a mixture of purple and yellow pansies at the west entrance to the campus. From past experience, Erasmus knows that fewer than 350 pansies will fit in the planting area. Which inequality describes the situation?
   F $x + y \geq 350$ **H** $x + y \leq 350$
   G $x + y > 350$ **J** $x + y < 350$

41

Reading Strategies
Use Graphic Aids

There are infinitely many solutions for linear inequalities. That's why the solutions are shown as a graph on a coordinate plane. There are four possible ways to draw the line and shade the correct half plane. These correspond to the four inequality symbols $>$, $\geq$, $<$, $\leq$.
Use the graphic aid below as a guide.

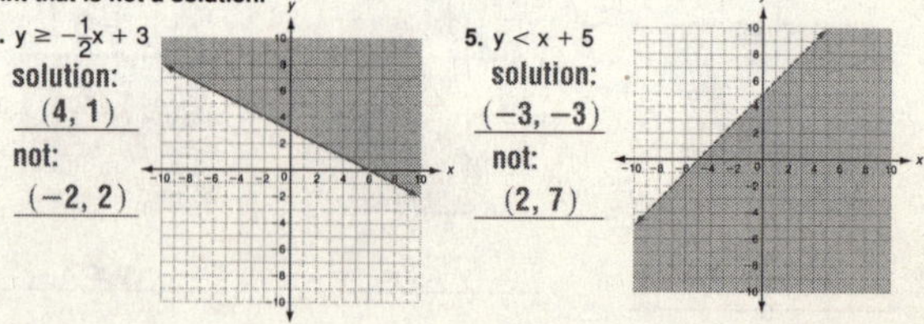

For each linear inequality, tell whether the graph is a solid or dotted line, and whether the shading is above or below the line.

1. $y \leq x + 7$

   solid;

   below

2. $y > -5x$

   dotted;

   above

3. $y \geq 3x - 2$

   solid;

   above

Graph each linear inequality. Name one point that is a solution, and one point that is not a solution.

4. $y \geq -\frac{1}{2}x + 3$
   solution:
   $(4, 1)$
   not:
   $(-2, 2)$

5. $y < x + 5$
   solution:
   $(-3, -3)$
   not:
   $(2, 7)$

42

Practice A
Solving Systems of Linear Inequalities

Tell whether the ordered pair is a solution of the given system.

1. $(4, 5)$; $\begin{cases} y \le x + 2 \\ y \ge x - 1 \end{cases}$
2. $(1, 3)$; $\begin{cases} y > 3x \\ y < x + 2 \end{cases}$
3. $(2, 3)$; $\begin{cases} y < 5x - 3 \\ y \ge -x \end{cases}$

__yes__ __no__ __yes__

Graph the system of linear inequalities. a. Give two ordered pairs that are solutions. b. Give two ordered pairs that are not solutions.

4. $\begin{cases} y \ge x + 1 \\ y \le -2x \end{cases}$
5. $\begin{cases} y < 2x + 4 \\ y > x - 1 \end{cases}$
6. $\begin{cases} y > -x \\ y > -x + 3 \end{cases}$

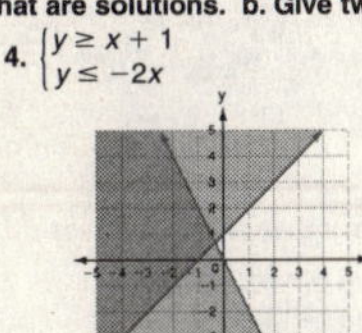 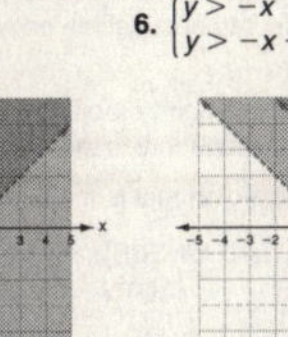

a. __$(-1, 0)$ and $(-3, 2)$__ a. __$(0, 0)$ and $(1, 2)$__ a. __$(3, 3)$ and $(4, 4)$__

b. __$(0, -3)$ and $(4, 0)$__ b. __$(1, 0)$ and $(-4, 3)$__ b. __$(0, 0)$ and $(0, 3)$__

7. Lou is buying macaroni salad and potato salad for a picnic. Macaroni salad costs $4 per pound and potato salad costs $2 per pound. Lou would like to buy at least 6 pounds of salads and wants to spend no more than $20.

   a. Write a system of linear inequalities.
   Let x = pounds of macaroni salad
   Let y = pounds of potato salad

   $\begin{cases} x + y \ge 6 \\ 4x + 2y \le 20 \end{cases}$

   b. Graph the solutions of the system.

   c. Describe all the possible combinations of pounds of salads that Lou could buy.

   __Any combination represented by the ordered pairs__

   __in the solution region.__

   d. List two possible combinations. __2 lbs mac. salad, 5 lbs potato salad;__
   __3 lbs mac. salad, 4 lbs potato salad__

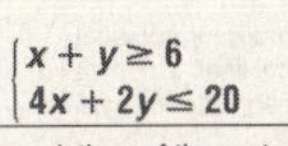

43 **Holt Algebra 1**

Practice B
Solving Systems of Linear Inequalities

Tell whether the ordered pair is a solution of the given system.

1. $(2, -2)$; $\begin{cases} y < x - 3 \\ y > -x + 1 \end{cases}$
2. $(2, 5)$; $\begin{cases} y > 2x \\ y \ge x + 2 \end{cases}$
3. $(1, 3)$; $\begin{cases} y \le x + 2 \\ y > 4x - 1 \end{cases}$

__no__ __yes__ __no__

Graph the system of linear inequalities. a. Give two ordered pairs that are solutions. b. Give two ordered pairs that are not solutions.

4. $\begin{cases} y \le x + 4 \\ y \ge -2x \end{cases}$
5. $\begin{cases} y \le \frac{1}{2}x + 1 \\ x + y < 3 \end{cases}$
6. $\begin{cases} y > x - 4 \\ y < x + 2 \end{cases}$

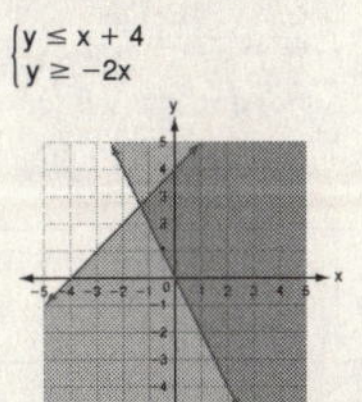

a. __$(0, 3)$ and $(3, -2)$__ a. __$(0, 0)$ and $(-2, 0)$__ a. __$(0, 0)$ and $(-2, -2)$__

b. __$(-2, 0)$ and $(-4, 3)$__ b. __$(3, 0)$ and $(2, 3)$__ b. __$(-3, 3)$ and $(4, 0)$__

7. Charlene makes $10 per hour babysitting and $5 per hour gardening. She wants to make at least $80 a week, but can work no more than 12 hours a week.

   a. Write a system of linear equations.

   x = babysitting hours,
   y = gardening hours,
   $\begin{cases} x + y \le 12 \\ 10x + 5y \ge 80 \end{cases}$

   b. Graph the solutions of the system.

   c. Describe all the possible combinations of hours that Charlene could work at each job.

   __Any combination of hours represented by the ordered pairs__

   __in the solution region.__

   d. List two possible combinations. __6 h babysitting, 4 h gardening;__
   __8 h babysitting, 2 h gardening__

44 **Holt Algebra 1**

Practice C
Solving Systems of Linear Inequalities

Tell whether the ordered pair is a solution of the given system.

1. $(-2, 3)$; $\begin{cases} y \le x + 5 \\ y > -2x - 1 \end{cases}$
2. $(-3, 3)$; $\begin{cases} y < -x + 1 \\ y > x - 4 \end{cases}$
3. $(-1, -2)$; $\begin{cases} y > x - 2 \\ y < 3x \end{cases}$

__no__ __yes__ __no__

Graph the system of linear inequalities. a. Give two ordered pairs that are solutions. b. Give two ordered pairs that are not solutions.

4. $\begin{cases} y \le 3x + 2 \\ y \ge -x \end{cases}$
5. $\begin{cases} y > \frac{1}{3}x - 2 \\ x + y > -3 \end{cases}$
6. $\begin{cases} 3x - 2y < 8 \\ y - 1 \ge \frac{3}{2}x \end{cases}$

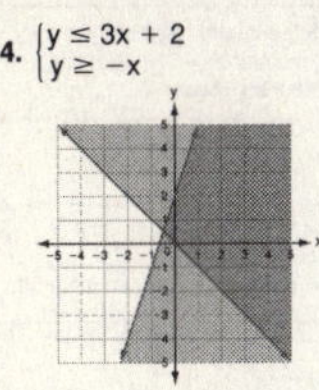 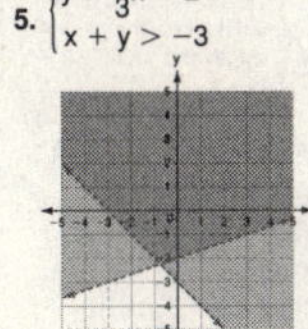

a. __$(0, 2)$ and $(3, 0)$__ a. __$(-2, 0)$ and $(0, 3)$__ a. __$(0, 1)$ and $(0, 0)$__

b. __$(0, -3)$ and $(-4, 0)$__ b. __$(-4, -2)$ and $(0, -4)$__ b. __$(4, -4)$ and $(0, -4)$__

7. Dennis works at a frozen yogurt store in the summer. He needs to order boxes of small cups and boxes of large cups. The storage room can hold up to 10 more boxes of cups. Each box of small cups costs $100 and each box of large cups costs $150. A maximum of $1200 is budgeted for cups.

   a. Write a system of linear equations.

   x = boxes of small cups,
   y = boxes of large cups,
   $\begin{cases} x + y \le 10 \\ 100x + 150y \le 1200 \end{cases}$

   b. Graph the solutions of the system.

   c. Describe all the possible combinations of boxes of cups that Dennis can order.

   __Any combination of cups represented by the whole number__

   __ordered pairs in the solution region.__

   d. List two possible combinations. __6 small, 4 large; 3 small, 5 large__

45 **Holt Algebra 1**

Reteach
Solving Systems of Linear Inequalities

You can graph a system of linear inequalities by combining the graphs of the inequalities.

Graph of $y \le 2x + 3$ Graph of $y > -x - 6$

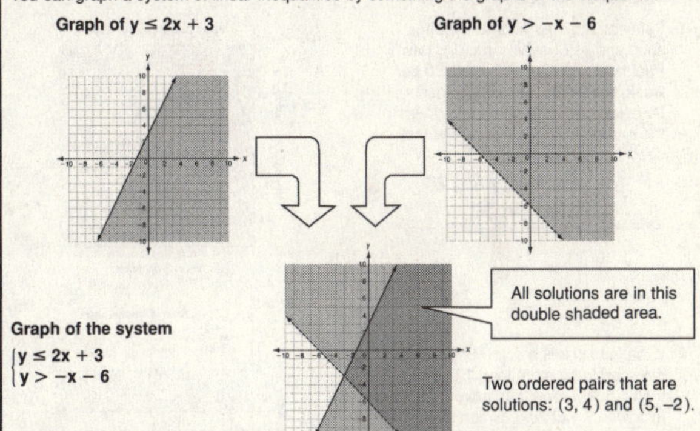

Graph of the system
$\begin{cases} y \le 2x + 3 \\ y > -x - 6 \end{cases}$

For each system below, give two ordered pairs that are solutions and two that are not solutions.

1.
2. Possible Answers:
Sol: $(1, -3)$, $(-2, -4)$
not sol: $(3, -3)$, $(-3, 2)$

Possible Answers:
Sol: $(-4, 3)$, $(-3, 1)$
not sol: $(3, -3)$, $(2, 2)$

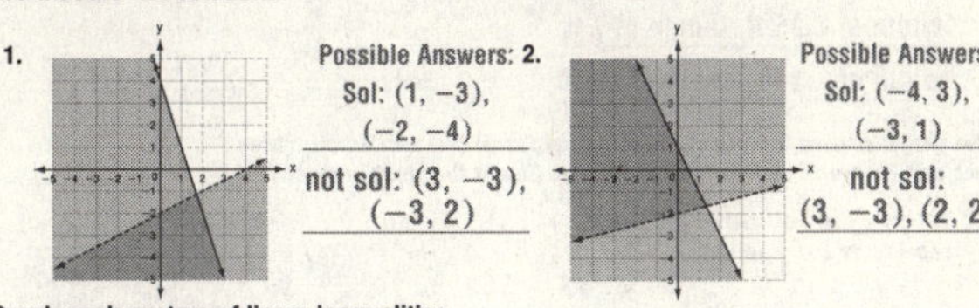

Graph each system of linear inequalities

3. $\begin{cases} y > x - 3 \\ y \ge -x + 6 \end{cases}$
4. $\begin{cases} y < x \\ y > -2x + 1 \end{cases}$

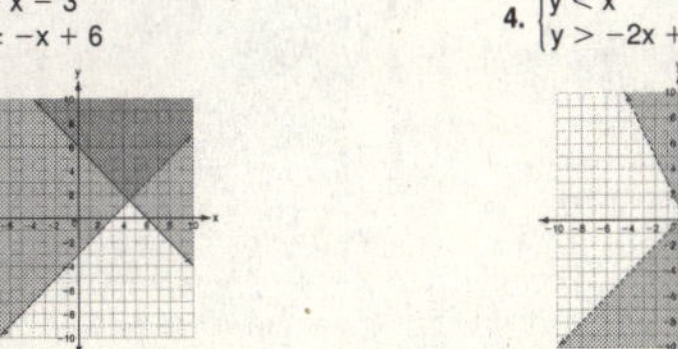 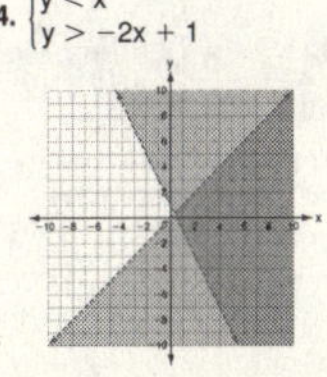

46 **Holt Algebra 1**

Reteach
Solving Systems of Linear Inequalities (continued)

A system of equations with parallel lines has no solutions.
Parallel lines in a system of inequalities might have solutions.

Graph $\begin{cases} y < x + 4 \\ y > x - 2 \end{cases}$.
Graph $\begin{cases} y \geq 2x + 4 \\ y > 2x - 1 \end{cases}$.
Graph $\begin{cases} y > -3x + 5 \\ y < -3x - 3 \end{cases}$.

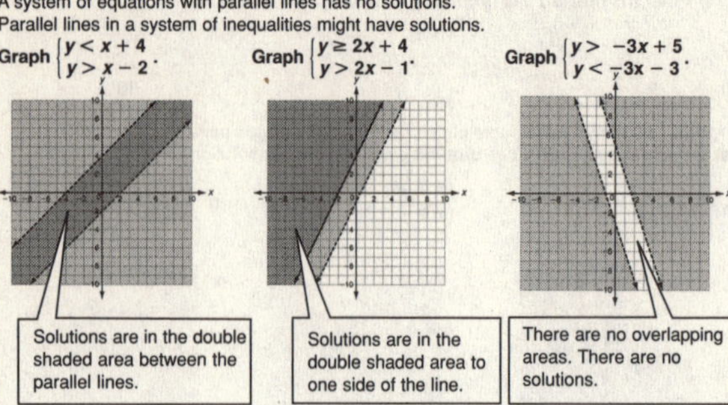

Solutions are in the double shaded area between the parallel lines.

Solutions are in the double shaded area to one side of the line.

There are no overlapping areas. There are no solutions.

Graph the solutions of each linear inequality.

5. $\begin{cases} y \leq x - 3 \\ y > x + 3 \end{cases}$

6. $\begin{cases} y > 2x - 2 \\ y \leq 2x + 3 \end{cases}$

7. $\begin{cases} y > -x - 1 \\ y > -x - 5 \end{cases}$

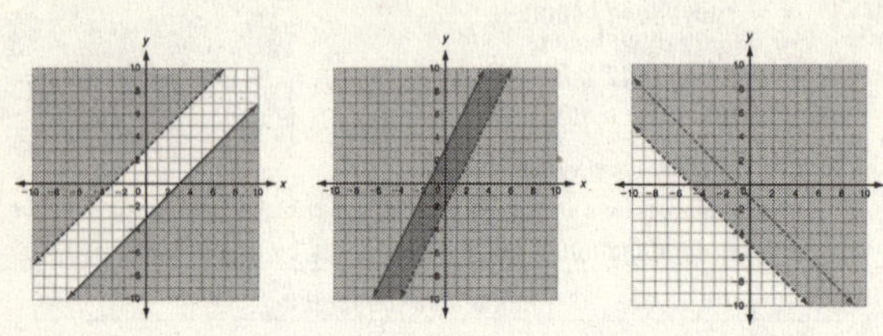

Holt Algebra 1

Challenge
Solving Systems of Linear Inequalities

Recall that $|x|$ has two parts in its definition. If $x \geq 0$, then $|x| = x$. If $x < 0$, then $|x| = -x$. This definition is useful when dealing with systems of inequalities that involve absolute value.

In Exercises 1 and 2, consider $\begin{cases} y \geq |x| - 4 \\ y \leq -|x| + 4 \end{cases}$.

1. Consider $y \geq |x| - 4$.

   a. On the grid at right, graph $y = |x| - 4$. Shade the part of the plane that is the solution to $y \geq |x| - 4$. (Test points to help decide on what part of the plane to shade.)

   b. On the grid at right, graph $y = -|x| + 4$. Shade that part of the plane that is the solution to $y \leq -|x| + 4$.

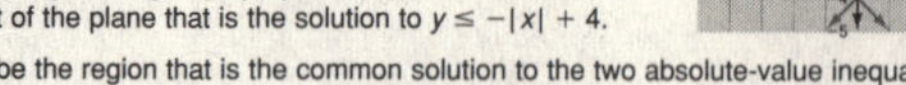

2. Describe the region that is the common solution to the two absolute-value inequalities.

The solution is the interior and boundary of the square formed by the absolute value inequalities.

You can reverse the process and represent a specified region by using a system of inequalities.

3. On the grid at right, graph the vertical lines $x = 3$ and $x = -3$. Also graph the horizontal lines $y = 5$ and $y = -5$.

   a. Write a pair of absolute-value inequalities whose graphs are the interior of the rectangle formed.

   $$|x| < 3, \, |y| < 5$$

   b. Write a pair of absolute-value inequalities whose graphs are the exterior of the rectangle formed.

   $$|x| > 3, \, |y| > 5$$

4. Explain how to modify the solution to Exercise 3 in order to represent the rectangle whose sides are bounded by the vertical lines $x = -1$ and $x = 3$ and the horizontal lines $y = -2$ and $y = 8$. Then write the system of absolute-value inequalities.

   Explanations may vary but should give the following absolute value inequalities: $|x - 1| \leq 2$ **and** $|y - 3| \leq 5$

5. The sides of a square are horizontal and vertical line segments whose diagonals meet at the origin. Represent the square and its interior by using a system of absolute-value inequalities.

   $$\begin{cases} |x| \leq a \\ |y| \leq a \end{cases}, \text{ where } a > 0$$

Holt Algebra 1

Problem Solving
Solving Systems of Linear Inequalities

Write the correct answer.

1. Paul earns $7 per hour at the bagel shop and $12 per hour mowing lawns. Paul needs to earn at least $120 per week, but he must work less than 30 hours per week. Write and graph the system of linear inequalities that describes this situation.

   $$\begin{cases} 7x + 12y \geq 120 \\ x + y < 30 \end{cases}$$

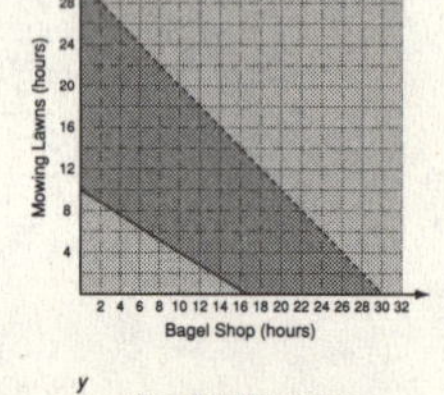

2. Zoe plans to knit a scarf. She wants the scarf to be more than 1 but less than 1.5 feet wide, and more than 6 but less than 8 feet long. Graph all possible dimensions of Zoe's scarf. List two possible combinations.

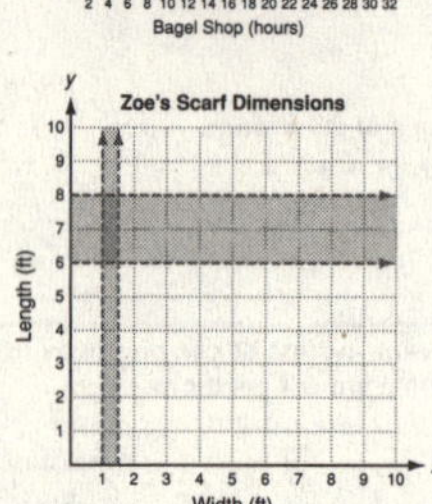

   width of 1.25 ft, length of 7 ft;

   width of 1.4 ft, length of 7.5 ft

The graph shows the numbers of two types of custom wood tables that can be made to fit a client's needs. Select the best answer.

3. Which system of linear inequalities represents the graph?

   Ⓐ $\begin{cases} x + y \leq 15 \\ y \geq 12 - \frac{4}{3}x \end{cases}$

   C $\begin{cases} x + y \geq 15 \\ y \geq \frac{4}{3}x - 12 \end{cases}$

   B $\begin{cases} y \leq x + 15 \\ y \geq 12 - \frac{4}{3}x \end{cases}$

   D $\begin{cases} y \leq 15 - x \\ y \leq \frac{4}{3}x - 12 \end{cases}$

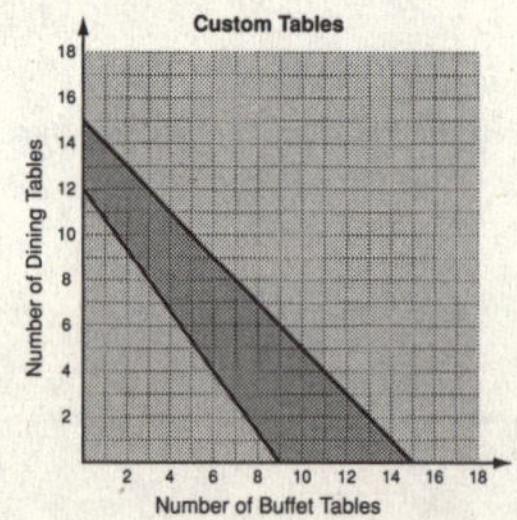

4. If 6 buffet tables are built, which can NOT be the number of dining tables built?

   F 4   H 8

   G 6   Ⓙ 10

Holt Algebra 1

Reading Strategies
Analyze Information

To solve a real-world problem involving inequalities, there is a lot of information that must be analyzed. Study the example below.

The perimeter of a garden must be less than 20 meters.
The width must be at least 6 meters.
Show all possible combinations of the garden's length and width.
Identify two possible solutions.

- Let x = length of garden
- Let y = width of garden

Define variables for the unknowns: "the garden's length and width."

The perimeter must be less than 20 meters: $P = 2l + 2w$; and "less than" means $<$.

$$\begin{cases} 2x + 2y < 20 \\ y \geq 6 \end{cases}$$

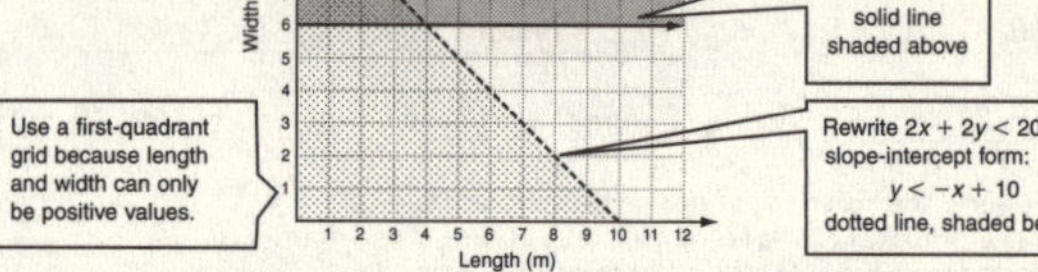

Solutions are all points in the double-shaded region, but not on the dotted line.

The width must be at least 6 meters; y = width and "at least" means $\geq$.

$y \geq 6$ solid line shaded above

Use a first-quadrant grid because length and width can only be positive values.

Rewrite $2x + 2y < 20$ in slope-intercept form: $y < -x + 10$ dotted line, shaded below

Possible solutions: $l = 2$ m, $w = 7$ m and $l = 1$ m, $w = 8$ m

Manuel is going to build a rabbit cage with a rectangular base.
The perimeter can be no greater than 30 feet.
The length must be greater than 8 feet.
Let x = length and y = width.

1. Write a system of linear inequalities to describe the possible cage sizes.

   $$\begin{cases} 2x + 2y \leq 30 \\ x > 8 \end{cases}$$

2. Graph the system to show possible dimensions.

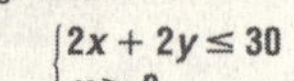

3. Give two possible dimensions for the cage.

   $l = 10$ ft, $w = 5$ ft and

   $l = 11$ ft, $w = 4$ ft

Holt Algebra 1

Holt Algebra 1